National Energy Issues — How Do We Decide?

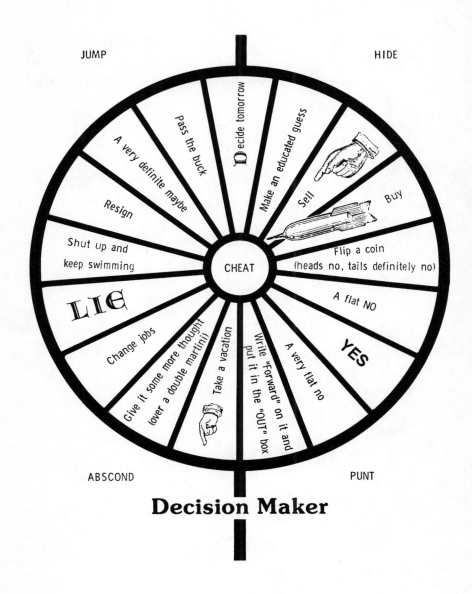

JUMP   HIDE

Decide tomorrow
Make an educated guess
Pass the buck
A very definite maybe
Resign
Sell
Buy
Shut up and keep swimming
CHEAT
Flip a coin (heads no, tails definitely no)
LIE
A flat NO
Change jobs
YES
Give it some more thought (over a double martini)
Take a vacation
Write "Forward" on it and put it in the "OUT" box
A very flat no

ABSCOND   PUNT

Decision Maker

National Energy Issues—How Do We Decide?

Plutonium as a Test Case

edited by
Robert G. Sachs

Proceedings of a Symposium and Post-Symposium Dialogue
of the
American Academy of Arts and Sciences/
Argonne National Laboratory

Ballinger Publishing Company ● Cambridge, Massachusetts
A Subsidiary of Harper & Row, Publishers, Inc.

Gift

 This book is printed on recycled paper.

International Standard Book Number ISBN 0-88410-620-9

Library of Congress Catalog Card Number: 79-18341

Printed in the United States of America

Library of Congress Cataloging in Publication Data

Main entry under title:

National energy issues—how do we decide?

   Includes bibliographical references.
   1. Energy policy—United States—Congresses.
2. Nuclear fuels—Congresses. 3. Plutonium—Congresses.
4. Atomic energy industries—United States—Safety measures—
Congresses. I. Sachs, Robert Green, 1916- II. American
Academy of Arts and Sciences, Boston. III. United States.
Argonne National Laboratory, Lemont, Ill.
HD9502.U52N35   333.7   79-18341
ISBN 0-88410-620-9

Contents

Acknowledgments

This symposium served as a focus for the 1978 Stated Meeting of the Midwest Center of the American Academy of Arts and Sciences held at Argonne National Laboratory. The choice of topic—national energy issues—was a natural one for the location since Argonne is a major national resource for the technical development of energy options as well as methods of controlling and understanding the impact of energy production. Furthermore, consideration of plutonium as a test case was particularly appropriate because of Argonne's role in nuclear power development, especially the development of the Liquid Metal Fast Breeder Reactor.

The symposium was made possible by the sponsorship of the Midwest Center of the American Academy of Arts and Sciences and by the financial support of the Argonne Universities Association. Discussion with the Council of the Midwest Center, chaired by Harrison Shull, helped us to develop ideas about the content and format of the symposium. David Easton and William McNeill in a local advisory capacity were especially supportive, and William Sewell made a special, last-minute effort to help round up the excellent group of participants. John Voss, Executive Officer of the Academy, helped to get our plans and arrangements started. I am especially grateful for the continuing contributions of Corinne Schelling, Assistant Executive Officer, for discussions of content, for obtaining suggestions of names of prospective participants, for reading and commenting on my contribution, and for generally providing principal liaison with the Academy.

Patricia Failla of Argonne National Laboratory, who served as Coordinator, was of the greatest assistance to me throughout the proposal writing, planning, and executing stages of the symposium. The arrangements worked well because of the usual efficiency of Miriam Holden, Argonne's Conference Administrator and her staff. Much credit for the production of these proceedings is due to Elaine L. Novey, who did what was necessary to persuade the participants to turn in corrected manuscripts, edited them, and took care of many of the technical details of production.

R.G.S.

National Energy Policy —
A View from the Underside

Robert G. Sachs

The formulation of current national energy policy appears to have begun in the councils of the Joint Committee on Atomic Energy (JCAE) during the 1950s. The JCAE deliberations led to the 1962 Report to the President on Energy Sources that warned that our oil resources would soon be exhausted. Although the administration responded to this in various ways through actions of the Atomic Energy Commission, the first direct statement of official policy was President Nixon's First Energy Message on June 4, 1971. Elaboration of that statement was provided by the President's Second and Third Energy Messages, delivered on April 18 and June 29, 1973. The urgency of these messages was due to concern about dwindling world oil supplies and was reinforced by the oil embargo of October 1973.

The June 29 message had a substantial political content since it proposed the dismantling of the AEC and the formation of a Department of Energy and Natural Resources to carry out energy research, development, and demonstration, and a Nuclear Regulatory Commission to take over the regulatory functions of the AEC. The main thrust of all three messages was technological: the emphasis was on advanced nuclear fission systems, clean combustion of coal, coal liquefaction and gasification, nuclear fusion energy, and advanced solar technologies, although some attention was given to institutional measures directed toward conservation and increasing energy resources. Highest priority was given to the liquid metal fast breeder reactor (LMFBR) as the most advanced renewable resource. The LMFBR continued to be the cornerstone of official policy until the end of 1977, when the new administration revised national energy priorities.

By the time the revised policies started to take shape with the formation of the Department of Energy, the emphasis had largely shifted from technological issues to institutional and political issues. Already it had become clear, even to those of us involved in energy research and development, that science and technology could not provide all of the answers to questions concerning energy policy.

The considerations that became prominent in the energy debate included, in addition to straightforward questions of engineering economics, such broader issues as regulatory policy, economic philosophy, national and international behavioral patterns, politics, and ethics. This broadening of the issues introduced into the decision-making process different and often contradictory perspectives, which arise from differences in methodology and information as well as from differences of values and goals. The groups now participating in decisionmaking include, in addition to physical scientists and engineers, economists, sociologists, political scientists, representatives of energy-related industries, a variety of other special-interest groups, the news media, and the politicians who in the end must make the decisions. The approach to energy policy from these diverse viewpoints is much more difficult than it was when the only question seemed to be a choice among technologies. Faulty though they may have been, there were well-established procedures for making those narrower choices in a "rational" way, but now the question of whether the broader issues can be resolved rationally is itself at issue.

The difficulties associated with bringing together such diversity were described by Tjalling Koopmans in his presidential address to the American Economic Association. Speaking of his experience as a member of interdisciplinary advisory committees, he said:

> our interdisciplinary group soon finds that its diverse participants ask different questions; use different concepts; use different terms for the same concept and the same term with different meanings; explicitly or implicitly make different assumptions; and perceive different opportunities for empirical verification—which may lead them to apply different methods to that end. The result can be politely concealed bewilderment, possibly a suppressed surge of "we-and-they" feeling, in the worst case a growing mistrust that only time and sustained interaction can overcome. (1979:1)

Some of the reports emerging from interdisciplinary committees read as though they originated in the Tower of Babel.

In dealing with energy policy matters, we are faced with confusion and misunderstanding as a result of the conceptual mismatch between different groups and individuals. For example, while there is supposed to be a collective effort to make predictions, or at least projections, about energy resources and uses based on technological, economic, and social considerations, the concept of predictability in physical science differs greatly from that in economics or sociology. Although a quotation from Niels Bohr is often used to illustrate the problems involved in making predictions, namely, "it is very difficult to make an accurate prediction, especially about the future," Bohr, like other physical scientists, had confidence in the predictive power of the established physical laws. I do not know the context of the quotation, but if it was about physical science at all it must have been about its direction or the sociology of science. Bohr surely believed in the ability of the Vatican Astronomers to use calculations of planetary motion to predict the day on which Easter will fall far in the future.

When a physical scientist speaks of a prediction about a physical process he refers to a quantitative statement subject to test by repeated measurements under controlled conditions. For the prediction to be judged successful these measurements must invariably lead to the expected answer within the range of the directly determined random errors and the estimated systematic errors. Development of the art of minimizing systematic error is one of the great accomplishments of experimental science.

Economists make measurable predictions too, but, at least as perceived by a physical scientist, their idea of controlled repetition of an experiment is a different one—one that makes it very difficult to determine random errors. For them the determination of systematic errors in the sense of physical science probably would be a meaningless exercise. The behavioral scientist appears to be one step further removed from the concept of comparing measurements with predictions.[a] In fact, such considerations may not be important to discussions of behavioral patterns, although there is always the question of how to determine in advance whether either of two differing views of the future is the correct one.

Further the judgments of each of us are strongly affected by our ability to distinguish solid evidence from personal opinion. Our values and goals are reflected in our opinions, as are our professional

[a]The difficulties of physical scientists in confronting behavioral scientists are shared by others as illustrated by the experience of the 1973 delegation of scholars sent by the Committee on Scholarly Communication with the People's Republic of China to discuss scholarly exchanges. At a meeting with Chou En-Lai, one of the first things he did after looking over our roster was to ask: "Tell me, please, just *what* is a behavioral scientist?"

training and disciplinary practices. But people differ even about what constitutes an opinion, as illustrated by the following incident:

A congressman who had been persuaded to favor a "soft-technology" approach to the energy needs of the United States was discussing projections of demand for electricity with an engineer. He pointed out that with adequate conservation measures, a little modesty in our life-style, and other steps emphasized by the soft-technologists, the rate of growth of electrical demand could easily be held down to 2 percent per year. "But Mr. Congressman, even at the low rate of 2 percent per year, electrical demand will double in thirty-five years," said the engineer. "That's *your* opinion!" replied the congressman.

Although this story represents an extreme case, it does demonstrate how difficult it is to distinguish between what is opinion and what is evidence. The concept of evidence itself takes on different meanings for scientists and for attorneys. Since much of the process and many of the tactics of the energy debate are in the hands of politician-lawyers, that difference cannot be ignored. In the natural sciences evidence is the weight of all the reproducible information, including the possibilities associated with contrary evidence. In the courts or in debate, attorneys weigh evidence for its tactical value, and contrary evidence is avoided or discredited when possible.

We have already taken note of the differences in the values of the various parties to the energy debate. But it should be noted also that even the concept of value may have a different meaning for each of us. There have been many attempts to weigh the values of one group versus those of another (see, for example, Tribe, Schelling, and Voss 1976), but any measure, such as measures in terms of dollars and cents, will invariably stir controversy. Other measures, such as votes, or the results of an opinion poll, may be attempted, but none is likely to be generally accepted as a true measure of worth. Furthermore, judgments by some of the groups participating in the energy debate will undoubtedly continue to be made on the basis of visceral reaction, rather than objective or quantitative information. This is illustrated by the following policy statement proposed (but not adopted) by a committee of the National Council of Churches: "The underlying cause of the energy crisis is not natural forces, but sin. Sin is rebellion against God. When worship of God is replaced by faith in human ability to command and direct energy flows, humanity has fallen into the sin of idolatry."

A final and pointed example of a difference in interpretation is the word "problem." Engineers are "problemsolvers" and both engineers and scientists speak of a "good" problem. To them the word has, on the whole, a positive connotation. But to most people the word problem signifies that something has gone amiss, which therefore cannot be "good." Technologists seek a solution to a problem, sociologists try to understand the problems of our society, and politicians seek an accommodation to a problem. If decisionmaking were only an exercise in problemsolving, the technicians would prevail, but the need for a consensus on such an all-encompassing issue as energy means that the methods of politics must dominate in the end.

Given these complexities, what we as scholars can do is what we do best: try to understand the issues, seek solutions or accommodations within our own disciplines, and, above all, try to understand one another. The difficulty in attaining even these limited goals is evident in the strong disagreements on energy policy matters among conscientious scholars. Positions in total disagreement with each other are supported by what may appear to an outsider to be equally persuasive arguments. One day the newspapers report that a study has shown us to be on the brink of energy disaster, and the next day an equally authoritative study is reported to conclude that there is no problem. It is not surprising that under these circumstances the public and the decisionmakers are confused. The failure to communicate among those who are concerned with basic questions of meaning and knowing—the scholars, including scientists, engineers, economists, and sociologists, who are expected to be the source of wisdom for the public and government—seriously impedes the process of providing the public and polity with the tools for decisionmaking.

The symposium at which the papers in this book were presented was intended to be a modest step in the direction of improved communication among the participants in energy policy discussions with a background in the scholarly disciplines. It was thought that a brief exposure to the methodological and philosophical differences among disciplines might open the way for better communication and further inquiry into the process of decisionmaking in the energy field. In view of the enormous import for the future of decisions now being made about energy, any clarification of the decisionmaking process not only will be of value to future scholars, but also may benefit the nation and the world. There is always the remote possibility of improving a process if we understand it better.

Because there have been so many symposia, workshops, institutes, meetings, and other forums on energy policy matters in recent years, it is important to emphasize that this symposium was not intended to be just one more conference on what to do about energy. Participants were not asked to come prepared to present personal solutions to specific energy policy problems, although the organizers were well aware of the fact that almost every participant had strong views as to which are the "correct" and which the "incorrect" solutions. Instead they were asked to concentrate on the criteria they had used, or thought should be used, to arrive at solutions.

Establishing clearly formulated criteria for a decision would appear to be a logical starting point for arriving at the decision. It is not always apparent that decisionmakers have this in mind; indeed it seems likely that despite our commitment to rationalism, many of us have come to decisions on energy policy without having consciously formulated criteria. In focusing the symposium on criteria, we had the opportunity to illuminate the differences between making energy policy decisions on a rational basis and rationalizing decisions that have been made for emotional or other reasons. The participants were thus given an opportunity to clarify their own methods for making judgments. The papers of those participants who took up this challenge directly address criteria for decisionmaking explicitly. Although others were less explicit, it is possible to extract information concerning their criteria from their contributions without too much effort. Of course, there were also those to whom the question did not appear to be important. Nevertheless this symposium met with a modicum of success in attaining one of its objectives: to provide a forum for communication about criteria for making decisions on energy policy among scholars of diverse disciplinary backgrounds and personal views. Whether this will be followed by further communication or by further scholarly investigation of the methodological and philosophical questions remains to be seen. I can only hope so.

In trying to decide on the specific subject matter of the symposium the organizers found themselves in a quandary because of the richness of the material at their disposal. A number of possibilities were considered before the subject of plutonium was selected. Among the issues first considered were the role of science and technology versus the role of economics in energy policy, and, what turns out to be a related question, the role of government versus the role of industry in technological innovation.

Both technical and economic feasibility are imperatives for any useful energy policy, but there can be sharp disagreement as to whether technological innovation or market innovation are to be the driving forces on policy. If a new technology has a high probability of success in the marketplace, it can be expected that it will attract private investment. However, most of the current energy policy issues concern technologies with, at best, a long lead time for marketing and, at worst, a high probability for failure to meet tests of technical feasibility, economic feasibility, or public acceptance (political feasibility). Since it is in the public interest to assure adequacy of our long-term energy supply, it falls upon our government to buy the necessary "insurance" by funding development of at least some of the technical alternatives. This raises at least two questions: which alternatives, and how much to pay?

In dealing with these questions consideration must be given to all identifiable costs and benefits, which is a large order. Yet another question associated with costs is often overlooked: can every technical problem be solved simply by paying enough? As a result of past successes it is a common assumption that the answer is in the affirmative: "All we need is an Apollo program directed to solving the energy problem." This simplistic notion overlooks a range of economic, technical and social questions. In addition, because the government was the only customer for the Apollo Project, commercial feasibility was not at issue. Of course, if the objective of the Project had been to provide a paying tourist service to the moon, it would have provided a useful prototype.[b]

Discussions of how much the government should pay for the development of energy technologies usually relate the price too closely to what priority has been assigned a given technology. Generous budget allocations are perceived as indicators of high government priority. However, technical feasibility and technical sophistication bear a strong relation to cost, so that for the development of two alternative technologies appearing to have equal promise as sources of energy, there may be good reason to budget different amounts. For example, the cost of developing a "far-out" technology like nuclear fusion will be high compared to that of the further development of an established technology like coal liquefaction.

Another aspect of the relation between technical feasibility and cost is illustrated by solar energy development. There is no question that the source of energy is there and that it holds great promise for future space heating and cooling and for industrial and agricultural process heat. But there is one technical fact that cannot be altered by any extension of the technology at any price, because it is a fact of nature: solar energy arrives at the earth in a very diffuse form. The engineering and environmental costs associated with collecting it are bound to be large in any attempt to use solar energy as a central power supply; however, technical research and development can reduce both those costs and the front-end capital costs for decentral-

---

[b]A study carried out by Arthur D. Little, Inc. (1976) for the U.S. Department of Commerce concludes that the government has not usually been successful in its attempts at initiating innovation for the civilian market. An exception is nuclear power.

ized systems. Thus, even including the cost of research on the related economic, institutional, and environmental questions, the development of solar will cost less than research and development of a new, complicated technology such as fusion.

The notion that government funding of a program is the principal measure of its priority has led to some questionable decisions about high priority technologies that require relatively low-cost research and development. Thus, the pressure to give a "proper" weight to the priorities has led to a shift in emphasis from the inexpensive research and development to the very much more expensive large-scale demonstration facilities for central solar electric power, coal liquefaction, coal gasification, and so forth. Although a demonstration plant can be helpful in establishing commercial feasibility, there is a question as to whether that is a proper object for government spending.

Another very high-priority energy policy matter for which the correlation between priority and government spending is of doubtful validity is that of conservation. Although there are technological opportunities in this field, they tend to be based on state-of-the-art technologies rather than "high" technology, except for very long-term possibilities in the biological sciences. There are, however, extensive opportunities for research in economics and the social sciences, where there is a need for testing models by means of demonstration. But by the very nature of these demonstrations, the financial need will be for front-end capital rather than outright funding.

Notwithstanding its desirability, conservation is controversial because of differences in the way the term is interpreted. Once again, the interpretation is greatly influenced by the values and goals of the interpreter. There is an inclination of the industrial sector to view conservation as an attempt to reduce production and therefore an attack on the economic system. But conservation can mean just the opposite: increased productivity resulting from increased efficiency in energy use by both industrial plants and individual consumers.

From at least the time of J. Willard Gibbs it has been known that the useful work available from a given energy source is determined by thermodynamic considerations; that is, the laws of nature place limits on the efficiency to be obtained. But as long as those limits are not reached, there is a place for technological development in the conservation program. At the same time it is easy to identify institutional changes that should lead to greater efficiency. In order to evaluate such changes, such as the often proposed decentralization of society, it is necessary to consider the economic and sociodemographic consequences of any proposed action. These development efforts and evaluations call for government-supported investigations, but not at a cost comparable to that of a high-technology development. The issue is "How much is enough?" not "How much can we spend?"

With such an abundance of questions relating to decisionmaking on energy, the symposium-planners knew that it was necessary to limit the program to a relatively small slice of such a rich menu. One of the several possibilities we considered was to concentrate on the nuclear fuel supply, because current national energy policy assigns an important role to conventional nuclear power using the once-through fuel cycle. The success of this policy depends on the reliability of estimates of the uranium supply, about which there are both geological and economic views, not to speak of technical questions concerning the processing and quality of the ore. There are also questions concerning methods for artificially enlarging the fuel supply by means of the LMFBR or other advanced systems.

This topic leads directly to the subject of the plutonium fuel cycle, which became the "test case" of the symposium. It was decided to concentrate on this subject because of its controversial nature, because the extremes of policy positions were strongly represented, and because of its importance to the international as well as the national community. Furthermore, the topic seemed to cover nearly the full range of energy policy considerations, including technical, economic, social, political, and ethical factors.

This range is well illustrated by the titles of the four sessions: Decision Criteria for Needs for Nuclear Fuel; Decision Criteria for Health, Safety, and Environment; Decision Criteria for Nonproliferation Requirements; and The Decisionmaking Process and the Plutonium Question.

The first session, on needs for nuclear fuel, provided an opportunity to hear various points of view about energy supply and demand, especially for electrical energy. Since nuclear power is primarily a source of electricity, an estimate of fuel supply is only meaningful in the context of a knowledge of the supply of and demand for electricity. The session turned out to be a very stimulating meeting of people, if not of minds, illustrating just how great is the disparity in basic assumptions or philosophy with which we must deal in arriving at a decision.

The second session, on health, safety, and the environment, was largely devoted to rather careful statements about the basis for assessing the hazards of plutonium in particular and nuclear energy in general. Its substantial technical content serves to illustrate just how complicated these problems are. The question of criteria to be used in interpreting engineering safety analysis was addressed directly. However, it became clear that the sociopolitical approach is not amenable to such definitive analysis, since the notion of acceptable risk has more to do with what people perceive and want than with any precise engineering equations.

The other major issue raised by plutonium—nonproliferation—was taken up in the third session. The proliferation dangers associated with nuclear power, especially with the reprocessing of spent fuel to recover plutonium, were discussed, as were some ways to mitigate them. An analysis of the impact on electrical energy supply of various ways of extending our nuclear resources into the future suggested reasons for keeping open the fast breeder option. At the same time the momentous economic and social implications for the world of our policies and decisions were brought out in the discussion.

In the last session various approaches to studying decision-making, including "institutional engineering" and operations analysis, were outlined. These were then analyzed by the discussants—a political scientist, a psychologist, and an anthropologist.

Each session was followed by lively discussion among the participants and members of the audience. These proceedings attempt to capture the flavor both of the talks and of the discussions. The entire symposium was videotaped and for those who would find them helpful, tapes of any session can be borrowed from the library of Argonne National Laboratory.

After the symposium, Alvin Weinberg submitted some comments giving his impressions of what he had heard, which were distributed to the participants. There ensued a continuation of the discussion, this time by mail, which is included here as an appendix.

POSTSCRIPT

It is important to call attention to the fact that this symposium took place some months before the accident at the Three Mile Island reactor. The views expressed in these proceedings have not been adjusted to take that experience into account. Much time will be required to collect reliable information concerning the accident, to analyze it, and to draw conclusions about its implications. It would not have been in the scholarly and thoughtful spirit characterizing this symposium to have asked the authors to make changes in their presentations without the benefit of the results of that lengthy process. Therefore the proceedings are to be viewed as a statement of opinions held at the time of the symposium.

REFERENCES

Arthur D. Little, Inc. 1976. "Federal Funding of Civilian Research and Development." U.S. Department of Commerce report PB-251.

Koopmans, Tjalling C. 1979. "Economics among the Sciences." *American Economic Review* 69, 1 (March): 1.

Tribe, L.H.; C.S. Schelling; and J. Voss, eds. 1976. *When Values Conflict.* Cambridge, Mass.: Ballinger Publishing Company.

Session I
Decision Criteria for Needs
for Nuclear Fuel

In focusing on plutonium as the test case for this symposium we are raising two questions: whether or not there is a need for nuclear power and, if so, whether or not the need can be met by nuclear fuels other than plutonium. In addressing these questions it is necessary to provide some measure of the amount of nuclear power needed or desired for the future. Since nuclear power is primarily a source of electricity, one point of departure is to estimate the level of electrical demand at which nuclear technology becomes necessary and then to attempt to establish criteria for deciding when or whether that level of demand will be exceeded.

Different views of how to arrive at such criteria make up the content of this session. The initial presentations represent a highly analytical econometric approach indicating a growing demand for electricity requiring nuclear energy, and a more qualitative no-growth approach, starting from the assumption that the level of consumption of electricity can be set so as to avoid the need for nuclear power. The socioeconomic and political factors associated with these two extreme points of view and alternative positions are then discussed. This session becomes a microcosm of the debate between advocates of growth and their opponents. Thus from the beginning the symposium meets the goal of providing a series of confrontations.

Opening Remarks by

Chairman: *Alvin M. Weinberg*

The issues that are being discussed in this symposium occupy that blurry area where science and policy meet. The scientific issues are themselves fuzzy. We are talking about the future; we are talking about world events; we are talking about imponderables. Some of us like to speak of issues of this sort as trans-scientific, with the idea that if you assign some special designation to these issues, the problems themselves will go away or be made easier. Of course, they will not be made easier; one has to confront these issues with whatever knowledge one has and try to make policies as best one can.

While I was thinking about what this symposium was all about, I realized that it is concerned with the subset of the general question, "Can nuclear energy be made acceptable?" As many of the speakers and discussants of the symposium unfold their views, we will see that there are some who, on the one hand, believe that it fundamentally *cannot* be made acceptable, and others who believe that it *can* be made acceptable. It will be very interesting to see the degree to which those who hold the most polarized of views can engage each other in civilized discussion. We must recognize what I call an "Energy War." I hope that the outcome of this symposium will help us forge peace treaties in this Energy War, rather than escalate the conflict.

Chapter 1

Planning for Energy Supply under Uncertainty

René H. Malès

The debate on the need for nuclear energy, particularly on the use of plutonium as a fuel and on the development of breeder reactors, has focused on many issues. Some arguments have turned on the question of how much uranium is available. Others have concentrated on the question of availability of alternative technologies or fuels, such as geothermal, solar in all its forms, oil, and natural gas. There have been diverse major issues in different forums: the usability of coal, the risk of proliferation or diversion, the need for energy, and even moral values. Frequently, these arguments have been so homogenized that it is very difficult to distinguish among the issues or to distinguish the questions from the answers. In this discussion, the focus will be restricted to a single problem, the need for energy. Rather than deciding exactly how much energy we will need in some future year, or debating what fraction of that energy will have to be delivered in electric form, let us focus on other aspects of the need for energy.

What factors are involved in planning for the future? There are the questions involved in dealing with uncertainty, in evaluating the cost of errors in planning, and in estimating the problems of long-run adjustment. Moreover, there are the problems involved when we shift from the way things have been done in the past. For example, today we are confronted with much higher energy prices, consciousness of the environment, and awareness of the importance of conservation. When we understand these factors, we can make a simple projection of the energy needs for the future and evaluate the inferences of that projection.

FACTORS INVOLVED IN PLANNING

Uncertainty

Lincoln Moses, head administrator of the Energy Information Administration of the Department of Energy, prefaced his first report to Congress with the comment, "There are no facts about the future." One need only look at a sampling of projections of total energy requirements to see the truth of this statement. The extent of differences between projections is shown by Figure 1–1, which contrasts consumption forecasts for the year 2000 with the current energy use of about 76 quadrillion Btu (quads) [in 1977]. Many more forecasts could be shown between the lowest one and the next highest. It has been suggested by some that showing only high estimates indicates a bias, and it is true that there are numerous estimates in the range of 100 to 120 quads, and even some estimates in the range of 60 to 100 quads. Although the estimates stretch all the way from a low of 60 quads, somewhat less than the amount used today, to a high of 160 quads, the high end of the more bullish estimates, planning for the higher range of forecasts is more important, as will be shown later.

Why these different estimates? What do they reflect? In many cases, they reflect different assumptions about the future—about economic growth, energy prices, technical changes that are taking place, technologies that will be available and in use by the year 2000, and about policy and regulation.

Figure 1–1. U.S. Energy Consumption Forecasts *(Year 2000)*

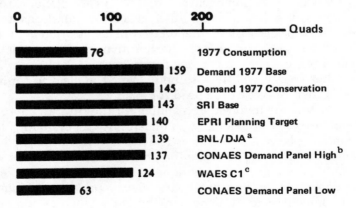

	Quads
76	1977 Consumption
159	Demand 1977 Base
145	Demand 1977 Conservation
143	SRI Base
140	EPRI Planning Target
139	BNL/DJA[a]
137	CONAES Demand Panel High[b]
124	WAES C1[c]
63	CONAES Demand Panel Low

[a]Brookhaven National Laboratory/Dale Jorgenson and Associates.
[b]Committee on Alternative Energy Systems, National Academy of Sciences.
[c]Workshop on Alternative Energy Strategies (MIT).

Another variable incorporated in these forecasts is the anticipation of what people will do or what they will want. Herman Daly (Chapter 2) focuses on the question of life-style changes and different societal objectives that he believes we should, or perhaps must, adopt. The low forecast of 63 quads, for example, assumes a significantly different life-style from that implied in the forecast of 150 to 155 quads.

A further difference is in estimates of the way the world will respond to stimuli. One determinant is the response of people to price increases or what economists term price elasticity. Another is the response to exhortations to change life-styles. Yet another is the response to new rules. For example, while the 55-mph speed limit has not been very effective in getting people to drive at 55 mph, it has lowered the average speed. New rules about energy use, about installing insulation, about maintaining one's automobile may not be followed to the letter of the law, but they will probably have a noticeable effect on living habits.

Another reason for the differences in the forecasts is that there are errors and inconsistencies in the measurement of energy consumption—both in the measurement of current use and in the prediction of future use. These involve the rules of accounting for solar energy, for other noncombustion energy (such as hydro and nuclear), and for losses along the energy production stream.

Uncertainties and differences of opinion about what is to happen in the future cannot be made to disappear. More information, better methods of analysis, clearer insights into human motivations can reduce the degree of uncertainty, but never to zero. The task is how to plan for the future despite uncertainty as to what the demand will be.

Nonsymmetry of Cost

The second factor we should consider is the cost of errors in planning and how to minimize the risk of errors. If we could anticipate the future perfectly, we could develop a plan for the expansion of the energy system that would result in the lowest possible cost of energy for consumers. Because we cannot know the future perfectly, we cannot plan perfectly. This defect in our knowledge incurs costs either from planning for more than we need or from planning for less than we need.

What are the costs of overplanning, and of underplanning? If we end up having slightly too much capacity, the cost is the carrying charge on the investment in that capacity, less the savings in efficiency, if any, from implementing new technology. This is shown in

Figure 1—2.   Relative Consumer Benefit

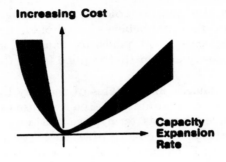

Figure 1—2 as the line sloping gently upward to the right, from the minimum cost or perfect planning point. The cost of too little capacity is considerably more complex. The initial cost of a shortfall represents a small amount of inefficiency in the system and is somewhat offset by the decreased carrying charge on the smaller investment. As the shortage becomes larger and larger, the cost reflects growing inconvenience, interruption of service, and losses associated with those interruptions. If the shortage is very large, we can expect large costs not only in production losses and in unemployment, but also in social stresses. These stresses could pit residential consumers against industrial consumers, the poor against the rich, one nation against another. There are already signs of such stresses, and the cause is not so much energy shortage as the cost of energy. For example, Long Island Lighting's time-of-day rates resulted in the industrial consumers taking court action to prevent what they viewed as an infringement of their rights by preferential treatment of residential consumers. Also, the old conflict between the poor and the rich can be seen in the debates on lifeline rates. The conflict between the OPEC nations and the importing nations has us teetering on the brink of economic disaster and possible armed conflict.

It is difficult, unfortunately, to quantify the costs of nonsupply or shortages in the total energy system. Classic measures of the cost of such shortages in electric supply exists, however, which may indicate what the dimensions may be for the total energy system.

The cost of a kilowatt-hour at the time these measures were being derived was on the order of $0.03 to $0.04. Economists estimated the cost of an unserved kilowatt-hour to be on the order of $0.60 to $1.50. Looking at the cost of outage to specific commercial customers, Ontario Hydro found a range as high as $5 to $20 per kilowatt-hour.

Several years ago the California Energy Commission solicited studies on the best approach to planning for the optimal rate of electrical system expansion. One contractor, SRI International, responded with a very interesting approach (Cazalet et al. 1976). SRI found that for a given energy projection, there is some minimum rate of expansion. Figure 1−3 shows this optimal rate. It also shows the same shape curve as the overplanning–underplanning generalized curve in Figure 1−2. If the systems are expanded faster than the optimal rate, there are some additional costs—the carrying charges on the excess capacity—but they are offset by the opportunity to get rid of some old oil-burning units. But if there is insufficient capacity, the costs calculated by SRI increase rapidly as the shortage increases.

In addition, the SRI study pointed out that if there is uncertainty in energy requirements, the optimal expansion plan shifts to the right. This shift reflects the need for additional capacity to cover uncertainty in the plan. The more uncertain the future load requirements, the larger the system expansion ought to be, as is shown by the second curve in Figure 1−3. This conclusion is obvious once it is demonstrated. If the uncertainty in the energy forecast is symmetrical and the cost of the shortage versus the overage is asymmetrical, then it is desirable to plan for some additional capacity to cover possible higher demands.

This idea appeared so fascinating to the Electric Power Research Institute that EPRI applied it to four test utilities, then expanded it to several other utilities. The methodology employed is essentially that of SRI, modified to carry out many more details of typical utility planning. Figure 1−4 shows the generalized results. Four types of cost are specified:

1. The environmental cost—the noninternalized impacts on the society from energy conversion of the electric system involved. Because reliable estimates of such costs are not available, they are assumed constant.

Figure 1−3. Effect of Uncertainty

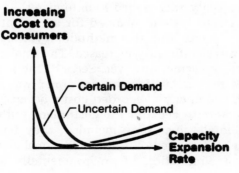

Figure 1-4. Cost to Consumers

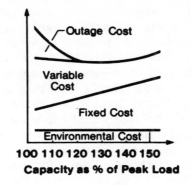

2. The fixed cost—primarily the carrying charge on the system investment.
3. The variable cost—operating and maintenance costs, and fuel costs which vary by the type of units used to meet demands. (The fixed and the variable costs are what the utility records on its accounting records, and what customers eventually pay for in their bills.)
4. The cost of outage incurred by the customer—such as lost production, social disruptions, or the loss of food in the refrigerator when the electricity goes off for a long time.

The Con Ed outage of 1976 was evaluated by the Congressional Research Service to have cost about $3 per kilowatt-hour. Some initial estimates ran to as much as $20 per kilowatt-hour when social disruptions were given very high values. In EPRI's study, these costs were evaluated in the range of $0.05 to $5 per kilowatt-hour.

The study resulted in a number of important conclusions. First, it showed that the larger the uncertainty about the load forecast, the more planned capacity was needed to minimize the total cost for the consumer. The tested systems planned for an operating reserve margin of 15 to 20 percent. With this methodology, the optimal planned reserve margin was significantly increased. The final operating reserve margin remained the same, however. Second, the study showed that the optimal reserve margin for the consumer was larger than that indicated by the company's recorded costs, because those recorded costs did not take into account the customers' outage costs. Third, when environmental values were included, they tended to increase reserve requirements; the reason is that additional capacity allows the decreased use or elimination of environmentally inefficient units.

Fourth, the study showed that for several of the systems tested, additional capacity lowered the cost the company incurred because of the company's ability to shift from using very expensive resources— namely, oil or natural gas—to coal or nuclear power.

The nonsymmetry of cost from overplanning versus cost from underplanning leads to two important conclusions: it highlights the value of planning for an adequate reserve, and it shows that uncertainty adds to the amount of reserve required to minimize customers' costs.

Long-Run Adjustment

The third factor involved in planning is the difference between short-run and long-run. Economists recognize that short-run supply curves are limited by the possible response. Figure 1–5 shows a series of hypothetical short-run supply curves that could be for the output of a given electrical system or the coal produced at a given point in time. The curves show that at any point in time, 1980 for example, the supply response is limited by the capacity in place. By 1985, response can be increased by putting in more capacity. The curve for 1985 is to the right, representing a greater quantity of output than the curve for 1980. The long run, the bottom curve in this figure, is traced by the optimal points of each individual annual supply curve.

It takes action to get from one short-run supply curve to another since it is not possible to jump smoothly over curves. For example, if after 1980 no productive facilities are added, we could not attain the 1985 supply curve shown in the figure by 1985. Even after 1985, if

Figure 1–5. Long-Run Supply versus Short-Run Supply

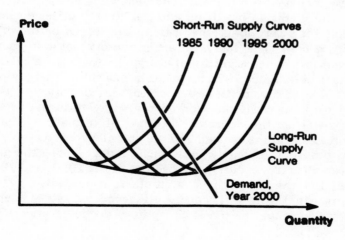

construction of energy facilities begins again, it is unlikely we could reach the 1990 supply curve plotted on this figure until 1995.

To be more specific, it now takes from 10 to 12 years to build a nuclear unit. It may take even longer if we do not simplify and rationalize the process of licensing. It now takes from 8 to 10 years to build a fossil unit; moreover, the time to build coal units is lengthening, and it is becoming increasingly difficult to obtain construction permits. More important, it takes from 20 to 30 years to go from the inception of a new technology to commercial availability. And that interval only produces its commercial availability; it does not provide the time for the technology to achieve a large market share. Similar investments of time are needed to develop fuel resources. To go from uranium believed to be in the ground at a specific location, to a fuel rod ready to insert into a nuclear reactor, takes 15 years. To convert coal from a known seam to fuel in the bunker takes 10 years. The oil in the Baltimore Canyon will take from 10 to 15 years to become a refined gallon of gasoline in somebody's gas pump ready to go into an automobile. The short run is different from the long run, but the long run is dependent on working through the short run, step by step.

ELEMENTS CAUSING CHANGE

Price

A number of important elements affecting the way our society operates have undergone or are undergoing significant changes. Over the period 1950–1973, energy prices actually declined at an average rate of 1.8 percent per year, the GNP increased by 3.7 percent, and energy use increased 3.5 percent. The increase in energy was primarily from the increased use of oil and gas. Energy prices declined or were stable for a very long time, a trend that began before this century (Figure 1–6). This has been an important factor in economic growth and in the increasing use of energy. So there is a chicken-and-egg element here. Since 1973 there has been a substantial jump in oil and natural gas prices and a lesser increase in coal prices. The price increases we have witnessed worked their way through the system before arriving at the consumer because of regulations, limits on price increases, and the averaging of old and new resources. However, although the price of energy will remain higher during the next 20 or 30 years than it was during the decades of the 1950s and 1960s, it is unlikely that the real price will continue to rise rapidly.

As shown in Figure 1–7, energy efficiency, or rather energy intensity, can increase or decrease in spite of declining prices. Energy

Figure 1–6. U.S. Energy Prices *(at point of production, 1975 dollars)*

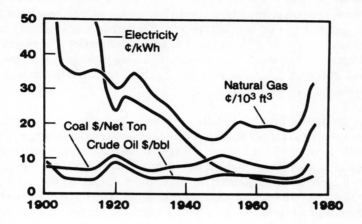

Figure 1–7. U.S. Index of Energy Consumption per Unit of GNP

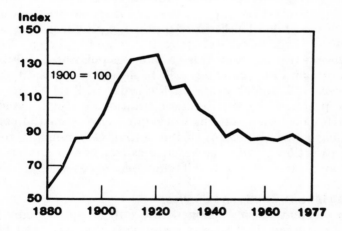

intensity declined precipitously during the 1920s and 1930s while the energy price was declining. Although analysts assumed that the decline in energy price was the causative factor, it reflected primarily the conversion from inefficient steam-driven systems to more efficient electrical ones. During the 1930s and 1940s, less efficient coal railroad engines were replaced by diesels, resulting in a transportation system that was more efficient although it used higher priced oil.

Conversely, energy intensity during the period 1880–1910 increased while prices were declining, reflecting a shift from an agricultural economy to an industrial one. The intensity change is overstated because energy use in agriculture is not recorded as carefully as it is in industry. If we were to project future consumption at the current level of energy intensity, the United States would be using between 180 and 200 quads by the year 2000. This is not likely for a number of reasons.

Environmental Energy Needs

One reason extrapolation from current levels of consumption does not work is we will need more energy to meet requirements for environmental protection. Significant inputs of energy are being used now to improve water quality in the United States, by such efforts as treating waste water, controlling chemical discharge, and reducing heat discharge. Energy is also being used to improve air quality by reducing emissions from automobiles and controlling emissions from stationary sources.

To get an idea of the extent of energy required, let us look at a few details. Current environmental protection standards for air quality require particulate control of emissions from electric power stations, which use about 1 percent of the energy produced by a power station. Nitrogen oxide controls may take about 4 percent. This is largely a guess since the standards are not yet published, and we are not sure what kind of technology will be necessary to meet them. A cooling tower will probably use on the order of 2 to 3 percent of the energy produced by the generating station. In total, standards envisioned for coal-fired stations will require environmental control systems using 10 to 15 percent of the output. Governmental regulations for protection in other areas, such as coal mining and offshore oil drilling and transport of oil, will require energy as well.

Changing Mix of Goods and Services

Another element causing change is the shift in our economy from goods or (manufacturing) to services, a shift that has been taking place since the 1920s. Although this shift has generally been associated with declining energy intensity, it is not clear that this association will continue. Some services are more energy-intensive than some manufacturing processes, so their respective shares of energy consumption depend on which services are becoming more prevalent and which area of manufacturing is declining.

Regulatory Shifts

Government regulation will also change our energy future. President Carter's National Energy Plan is geared to increasing energy efficiency and decreasing energy use, but it should be noted that not all its elements will act in that direction. For example, the coal conversion portion of the National Energy Act will tend to increase energy use.

There are other regulations that affect the energy future. The auto fuel efficiency regulation is intended to produce a much higher level of efficiency in transportation. Changes in local building codes and standards will reduce heating losses and make structures more energy efficient.

DERIVING A PLANNING TARGET

The future is not going to be like the past. What, then, is a simple and reasonable approach to planning? One way is to develop a simple model that explicitly incorporates the changing elements mentioned thus far. Although numbers are not as important as the ideas, they give us a calibration. We must also assume several important objectives. The first is to provide sufficient energy so that the availability of energy itself is not a constraint. The second is to provide for growth in the national economy, ensuring sufficient jobs for those who wish to work. The third is to ensure growth in productivity, providing for a rising standard of living. The fourth is to ensure that basic desired life-styles are available as options.

The model can be constructed in three steps:

1. Estimate GNP.
2. Derive gross energy requirements from the GNP estimate.
3. Modify the gross energy requirements for the changing elements we can perceive.

To estimate GNP, we need an estimate of the labor force, as is well known. The major uncertainties are the participation rate of the population and the current birth rate, which, it is interesting to note, are offsetting factors. If the birth rate is high, it is likely that the participation rate will be low; and if the participation rate is high, it implies a lower birth rate. In any case, it is assumed that the labor force will be almost 120 million by the year 2000. The average work week is projected to decline to 36 hours from the current 39 hours (Figure 1−8). Productivity is projected to grow at 2 percent per year

(Figure 1–9), resulting in a GNP of about $3.5 trillion (1975 dollars) by the year 2000.

The energy content of this projected GNP is derived from the historical relationship, which suggests a small improvement every year in the energy–GNP ratio, much as shown in the figure of the energy efficiency in the economy (Figure 1–10). These calculations imply a use of about 155 quads by the year 2000.

Energy requirements for environmental protection are expected to be very large. More detailed studies indicate that these might add about 16 quads to the total energy requirements.

Figure 1–8.   Work Week Trend

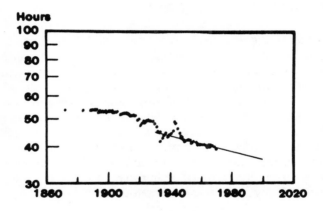

Figure 1–9.   Productivity

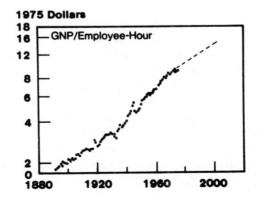

Figure 1—10. Energy versus GNP

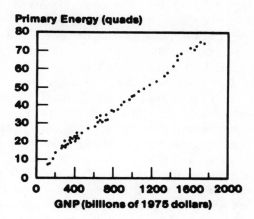

Conservation will be a counter factor. Higher energy prices are an inducement to conserve energy. Technology will allow the application of energy conserving techniques; regulation will induce changes. However, this will not require significant changes in life-styles. A study sponsored by EPRI found the potential for change, given current technology, to be about 40 percent of the total energy use, and it is estimated that only half this potential will be applied because of constraints and lags in the marketplace (see Table 1—1).

Manipulations like the foregoing result in a projected energy requirement of 140 quads by the year 2000, which is about 30 percent lower than that projected by simple extrapolation. It is interesting to note that this level of use is about comparable to the Swedish standards of energy intensity, adjusted for density of population, differences in climate, size of housing, and the mix between apartments and single-family dwellings. When econometric methods were used in

Table 1—1. Year 2000 Reasonable Energy Conservation Potential

	Sector Importance		Potential Energy Savings	Weighted Potential Annual Savings
	1975	*2000 (unperturbed)*		
Residential and commercial	35%	40%	15%	6%
Industrial	40	40	15	6
Transportation	25	20	40	8
Total	100%	100%		20%

an attempt to parallel the assumption of this projection, a separate econometric analysis found the same level of demand.

This projection is not an attempt to forecast the actual level of demand by the year 2000, but rather to develop a conservative target for energy supply planning. Such a planning target implies that if our society were prepared to meet such energy demands, no constraints would be imposed on the development of the society.

To translate the energy supply target to an electricity planning target, it is assumed that the growth in the electric share of energy market will continue at the present pace. The assumption reflects the effect of fuel availability and the constraints on the fuels that will be used. If the growth is in coal and in uranium, a larger fraction of the energy will be produced as electricity. This results in a projection of between 5 trillion and 9 trillion kilowatt-hours by the year 2000, with a midpoint planning base of 7 trillion kilowatt-hours. The United States used about 2.1 trillion kilowatt-hours in 1977, so that the rate of growth in electric energy is about 4 percent on the lower end and about 6.5 percent on the higher end; the midpoint growth is about 5.25 percent. All three are substantially lower than the 7.5 percent growth rate during the 1960s.

FUELS FOR GENERATION

The more pressing question is what fuels will be used to generate electricity (Figure 1−11). It is assumed that the amounts of natural gas and oil available will decrease and that significant amounts of coal liquids will not be available until after the year 2000. It is assumed also that maximum coal use is 1700 million tons per year. Currently, we are using about 480 million tons for production of electricity in the United States. There are two reasons it is prudent to plan on some sort of upper limit to the use of coal for electricity. Besides the problem of siting generating plants in such a way as to meet environmental protection standards, there is the possible competition for coal after the turn of the century, for use in synthetic liquids and gases.

The effects of conservation are expected to reduce significantly the prospective needs. Remaining, however, is a substantial need for nuclear capacity, for reprocessing the spent fuel, and eventually for the breeder reactor. Even then, there is a potential for shortfall sometime after the year 2000.

Figure 1−12 is a matrix of possible outcomes by the year 2010, based on how much uranium will be available to fuel reactors and on whether reprocessing and the breeder will be available. It shows the potential shortfall in meeting a planning target of 9.1 trillion kilo-

Figure 1–11. Electricity Generation Mix Projections

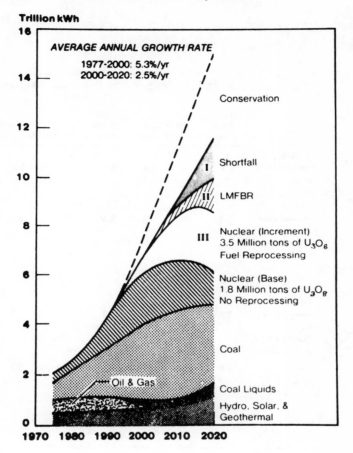

watt-hours. We can expect that the price of electricity will rise to dampen demand. Or we can believe that additional generation will be forthcoming from some type of crash program. For small differences between planned needs and planned capacity, it might be an acceptable risk to take. However, for very large shortfalls when options are not available, it is highly unlikely to be an acceptable risk. Does this prove that we need plutonium? Does this prove that we need a breeder? The answer is quite clear that it does *not prove* it. There exists no such proof, but it does illustrate the following.

Under reasonable but high projections of total energy and electric share, we could have a shortage of energy around the turn of the century, if extensive use of nuclear fuels is not possible. Because the consequences of a shortfall are costlier than the consequences of

Figure 1−12.   Electricity Sources for the Year 2010

	Hydro, Solar Geothermal	Oil/ Gas	Coal (base)	Light Water Reactor	LMFBR	Potential Shortfall	Total Generation
ELECTRICITY GENERATION BY INDICATED SOURCES IN 2010 (trillion kWh)							
Case A	0.8	—	3.4	1.5	—	3.4	9.1
Case B	0.8	—	3.4	2.3	0.2	2.4	9.1
Case C	0.8	—	3.4	3.2	—	1.7	9.1
Case D	0.8	—	3.4	4.5	0.2	0.2	9.1

Low (1.8 million tons) No — Fuel Reprocessing and LMFBR — Case A / Case B

U_3O_8 Reserves

High (3.5 million tons) Yes — Fuel Reprocessing and LMFBR — Case C / Case D

	U_3O_8 AVAILABLE (10^6 tons)	FUEL REPROCESSING
Case A	1.8	No
Case B	1.8	Yes
Case C	3.5	No
Case D	3.5	Yes

having too much capacity, because uncertainty leads to a need for greater preparedness, and because of the long lead times involved, we need to ensure the availability of these technology options if we are to meet future energy requirements. If the United States changes its objectives, if people change their life-styles, if the economy becomes substantially more efficient than we expect, the option does *not* have to be deployed just because it has been made available. But, the reverse is not true. If we do not obtain the levels of conservation, or life-style changes, or efficiency gains that we are projecting, and we have not prepared for that outcome and provided for employing more technologies, we force society into a mold it may not want. We foreclose the option of the higher level of growth. This is the price of not developing the options today. The price may be terribly high.

There has been much talk on the choice between two paths: the "hard technology" path and the "soft technology" path. Exponents of the soft technology path argue that development of hard technology options means foregoing development of soft technology. In fact, the reverse is true if we do not provide for the technical option of meeting the centralized energy requirements.

REFERENCE

Cazalet, E.; B. Judd; A. Miller; and E. Rasmussen. 1976. "Decision Analysis of California Electric Capacity Expansion." Prepared for the California Energy Resources Conservation and Development Commission, Stanford Research Institute, Menlo Park, California.

✳ *Chapter 2*

Decision Criteria for Needs
for Nuclear Fuel —
A No-Growth Perspective

Herman E. Daly

There are two independent sets of criteria for making decisions—possibility and desirability. Only alternatives meeting both criteria are eligible to be chosen. We must guard
against the twin errors of *wishful thinking* (assuming that whatever is
desirable must somehow, someday be possible), and *technical determinism* (assuming that whatever is possible must somehow, someday
be desirable).

Even within the intersection of the set of all things possible with
the set of all things desirable, there are many members. Some are
more desirable and more possible (that is, easier to achieve) than
others. To decide among desirable possibilities we must have some
measure of relative desirability and relative possibility. Economists
invent a measure of desirability and call it benefit (welfare, utility, a
sum of money). The measure for relative possibility is called cost and
is simply the benefit foregone—the benefit that would have been
yielded by the best alternative that has to be foregone when a particular alternative is chosen. This is the concept of opportunity cost,
often referred to nowadays as "trade-offs." A lower relative possibility is reflected in a higher opportunity cost. Next, we subtract the
cost from the benefit for each alternative to get net benefit. Then we
choose so as always to maximize net benefit.

In principle, making decisions about energy or anything else is so
easy! Why, then, do we hold conferences that agonizingly ask, "How
do we decide?" Of course, the simplicity of the abstract logic of
decisionmaking is very misleading since it proceeds in arithmomorphic fashion, reasoning with distinct, nonoverlapping concepts:

possible or impossible; desirable or undesirable; more or less desirable or possible in some quantitative sense, which allows arithmetic calculation of opportunity costs and maximization of a number representing net benefit. But, in fact, the possible and the impossible partially overlap, and the fuzzy boundary area itself shifts over time. Likewise, although logic does not allow something to be both desirable and undesirable, real life seems to be rather tolerant of such contradictions, as indeed it must be if there is to be evolution. If nothing could ever overlap and merge over time with its "other," we would have a nonevolving world that would be forever congruent with our static analytic concepts. But since we know that the world changes, we should not be surprised when evolving reality, like a jellyfish, slips through the static mesh of arithmomorphic concepts amenable to logic.

When we add the fact that we have no acceptable unit with which to measure benefit and cost, and consequently in which to express net benefit, as well as the fact that we have only the vaguest notion of what we really desire in the first place, then decisionmaking begins to appear impossible rather than easy. Nevertheless, we do make decisions, and it really is possible to reason with vaguely defined and partially overlapping concepts. Furthermore, such dialectical reasoning (see Georgescu-Roegen 1971: ch. 2), if done with good judgment, is far more important in life than analytical logic-chopping and number-crunching, because it more nearly corresponds with the reality that we try to map in our thinking. It is not for nothing that the brain has a right hemisphere as well as a left.

Economic growth and growth in energy use have over recent decades become both less beneficial and more costly. Although in particular instances the net benefit of further growth is no doubt positive, there are other instances in which net benefit is negative. Leaving aside the question of whether the aggregate balance comes out positive or negative, I suggest the hypothesis that in the case of plutonium-based energy, the net benefits are negative. It is only by adopting a very uncritical attitude toward the benefits of growth, that one can make a case for plutonium.

As an example of this uncritical attitude, let me cite a speech given in 1975 by a Nobel Prize-winning chemist, Dr. Willard Libby. After explaining in great detail all the dangers and hazards of plutonium, he concluded not that fission power was a bad bet, but that it was our only hope. A member of the audience raised the obvious question: "In view of the enormous risks and costs you have emphasized in such detail, what in your mind is the offsetting benefit that outweighs the cost?" The chemist drew an exponential curve on the blackboard and labeled it "energy demand." "Without fission power," he said, "we cannot keep up with demand." The questioner suggested that "energy demand" is merely a socioeconomic trend, not a constant of nature, like the speed of light, and that we do have the alternative of bending the curve downward, and if necessary, reducing it to a level that could be maintained without fission. The chemist felt that would imply mass unemployment, revolution, and economic chaos. The questioner felt that continued growth in a fragile and finite ecosystem would sooner or later imply ecological destruction, mass unemployment, revolution, and economic chaos—sooner, if plutonium were the basic energy source. The short run economic "imperative" of growth conflicts headlong with the long run ecological imperative of nongrowth. Which is the more intelligent resolution: to attempt to make the planet earth start to grow (presumably at a rate equal to the rate of interest), or to attempt to make the economy cease growing in its physical dimensions?

The very word "growth," as defined in Webster's dictionary, means "to spring up and develop *to maturity.*" Maturity is a point beyond which growth ceases and maintenance becomes the paramount material concern. Growth, in the metaphorical sense of intellectual or spiritual growth, is not at issue. No one opposes growth in that sense. The problem with the ideology of the economic growth imperative was pointed out years ago by another Nobel Prize-winning chemist, Frederick Soddy, who was also a good economist: "You cannot permanently pit an absurd human convention, such as the spontaneous increment of debt (compound interest), against the natural law of the spontaneous decrement of wealth (entropy)" (Soddy 1922: 30).

My task here, however, is not to make the case against growth, which has been done elsewhere (see, for example, Daly 1977), but rather to discuss from a no-growth perspective the decision criteria for needs for nuclear fuels. The word "needs" suggests that our concern is with criteria of desirability. The obvious question in defining needs is: "Needed for what?" Need implies a purpose. What purpose do we presuppose in speaking of need? What are our immediate purposes and what is the ultimate purpose or *causa finalis* from which they derive? Let us suppose that the latter is the well-being of mankind. How much energy is needed for the well-being of mankind? To know what is good for man, we must know something about the nature of man: what are we? Are we spirits temporarily tied to matter-energy; or are we matter-energy temporarily suffering from hallucinations—complex mechanisms with delusions of grandeur—or children of God with inferiority complexes? "Know thyself" is the first rule of philosophy, and lack of self-knowledge is our basic problem in determining criteria for energy needs. What we need depends on what we are. Yet we seem to assume that self-knowledge is impossible, or rather that since we are simply whatever our culture has made us, there can be no inherent nature of man to be discovered by self-knowledge, but merely a product of external forces that can be understood only by studying the external forces that molded us. When we think about the future, we ask only what we will become, not what we should become, because we lack the basis of independent selfhood and self-knowledge for doing anything other than projecting the trends which conditioned us thus far. Trend becomes destiny. Efficient cause usurps the place of final cause.

Projecting a mechanistically determined future and then treating it as a constant of nature to which we must adapt is a procedure unworthy of any organism with a central nervous system, much less with a cerebral cortex. To those whose self-knowledge includes a spiritual dimension, this approach is almost inconceivable in its total inversion of ends and means.

As J.K. Galbraith put it: "It is silly for grown men to concern themselves mightily with satisfying an appetite and close their eyes to the obvious and obtrusive question of whether the appetite is excessive" (1958: 98).

FOUR CRITERIA, TWO PATHS

Let us face up to the question "Needed for what?" at a more immediate level. Whatever the ultimate purpose, there are four proximate purposes or criteria that must be specified before it makes any sense to speak of energy needs. Energy is needed:

1. To maintain what size population (or what rate of growth)?
2. Living at what level of average per capita energy use (and how distributed)?
3. For how long a time period (indefinitely, or for 20 years)?
4. Using what kinds of technology (centralized nuclear electric, small-scale solar, or what)?

Any notion of future energy needs must give some general answer, either explicit or implicit, to each of these questions. The usual approach gives implicit answers in somewhat the following way. Recent growth rates of population and per capita energy use are projected forward to some round-numbered date. Whatever technologies are required to produce the projected amount are automatically accepted along with their usually unspecified social implications. Nothing at all is usually said about how long the system can continue beyond the target year. History either stops or starts afresh on the bimillenial year or the year 2050 or whatever.

Other answers to the four questions are at least conceivable. Since the four questions are quite interrelated, they cannot be answered independently. It would be unrealistic, for example, to choose larger population and higher per capita energy consumption, while at the same time opting for only soft technologies, and for sustainability over an indefinitely long time period. Likewise, if we choose soft technologies and a long life for the system, we must accept stabilized or lowered levels of population and per capita energy use. A further trade-off is that between population size and per capita use: many people at lower per capita use versus fewer people at higher per capita use. There is yet another constraint on the compatibility of goals, if we require that the combination we choose be generalizable to the world as a whole. It is clear to many that our current average energy use and industrial technological base are not generalizable to the total world population even for a short period of time, much less indefinitely, and still less so for a growing world population.

In choosing an optimal set of answers to these questions, two kinds of information are necessary. They are, first, technical information about the trade-offs among the four elements of purpose and, second, value information about what is good. The first kind of information allows us to determine which combinations of various levels of attainment of the four goals are in fact possible, and the second allows us to say which of the possible combinations is most desirable. Clearly, we need more research on *both* of these questions. But I think we know enough to make a broad decision between two general directions. Basically there are two general, self-consistent package answers to the four questions: (1) the high-demand, hard-technology answer, and (2) the low-demand, soft-technology answer.[a]

The hard energy package says that there is no such thing as enough, that growth in both population and per capita energy use is either desirable or inevitable; that it is useless to be concerned about the future more than twenty years ahead, since all reasonably discounted costs and benefits become small over that period. It says that the increase in scale, complexity, and centralization of technology is

---

[a] For a demonstration that there are alternatives to the "hard energy path," see *A Time to Choose* by the Ford Foundation's Energy Policy Project (1974). For elaboration of the soft path and for the most cogent analysis, see Amory Lovins' *Soft Energy Paths* (1977).

simply time's arrow of progress, and that refusal to follow it represents a failure of nerve, or doomsday pessimism.

The soft energy package gives far more weight to permanence and ecological stability, opting for technologies that not only use renewable sources of energy, but also are environmentally benign, less centralizing, less vulnerable to accident and sabotage, and controllable by the ordinary local people who depend on them, rather than by a remote priesthood of technical experts. The *rate* at which energy is available from such sources, however, is limited, and may require the reduction of per capita use to European levels, or about one-half of current levels (about equal to 1960 levels in the United States). Furthermore, population must be stabilized or reduced.

The contrast between the hard and soft energy futures can be made more dramatically explicit by considering the choice between the archetypal hard technology, nuclear fission, and the archetypal soft technology, small-scale solar energy. The question can be put in two parts as follows:

1. Do we want a small-to-medium scale, decentralized energy system subject to local control by the same people who use the energy; with minimal depletion or pollution and minimal disruption of ecological services; which arrives already distributed and is useless to saboteurs, terrorists, and psychopaths; which can be equally available to all future generations independently of our usage; which will be highly beneficial to poor tropical countries; and which, since the whole biosphere is preadapted to the fixed solar flux by millions of years of evolution, imposes upon us an automatic and healthy ecological discipline?

2. Or, do we want a plutonium-based economy, centrally administered by a technical elite; which is sure to permit the international and subnational proliferation of nuclear weapons; to increase our vulnerability to terrorists and to the inherent instabilities of large complex systems; which increases the dependence of poor countries on rich countries; which requires garrisonlike security measures at many points in the fuel cycle; which wastefully generates and distributes energy at too high a quality which must be wastefully degraded to fit the majority of end uses; which is based on highly uncertain assumptions about uranium supplies; and which imposes permanent absolute obligations on all future generations?

The contrast stated in the preceding two paragraphs will leave no one in doubt as to where I stand on the issue of fission versus solar energy and will lead some to dismiss my views as biased. But let us at least admit that the fact of having arrived at a definite conclusion is not by itself evidence of bias, any more than not having arrived at a conclusion is evidence of lack of bias (the latter may reflect only lack of thought). In advocating the soft energy path generally, we must guard against making absolutes of the relative virtues of smallness, decentralization, simplicity, and permanence. Some enterprises, such as insurance companies, cannot work on a neighborhood scale; the economies of factory organization cannot be taken advantage of below some minimal scale of demand; we cannot avoid *all* reliance on experts, short of limiting ourselves to those technologies understandable by the dumbest member of society. If we are to strive for infinite permanence, then we must never use any nonrenewable resource, and even that would not allow us to attain such an extreme goal; complete decentralization of all decisions to the individual or village level is made impossible by the existence of public goods and the tragedy of the commons.

Clearly, both the hard and soft energy paths can be reduced to absurdity by exaggeration. The difference is that while the reduction to absurdity of the soft path is an imaginative exercise, the reduction to absurdity of the hard path is being carried out in concrete, steel, copper, zirconium, and plutonium right before our eyes.

It seems to me that the values we profess, on ceremonial occasions at least, are much better served by redirecting our efforts to the soft energy path. Yet present policy has been, and to a lesser extent continues to be, to devote vast sums to breeder reactors, and comparatively little to solar energy and conservation. We continue to follow the hard energy path by default of enough imagination to recognize the alternative.

PRICE-DETERMINED VERSUS
PRICE-DETERMINING DECISIONS

The choice between a hard-path and a soft-path future is probably the major social and moral decision facing our generation. Treating it as an economic decision has obscured the issue greatly. Economics, "the sordid lore of nicely calculated less or more," is appropriate at the level of tactics, but not at the higher level of strategy. The choice between, say, oil and natural gas for some particular use is an economic choice; likewise, the choice between photovoltaic and biomass conversion. In the first case, the alternatives are both fossil fuels, and we are comparing one form of geocapital consumption with another. In the second case, both alternatives are solar income sources. But the choice between solar income sources and geocapital sources, say biomass versus natural gas, is more a question of long-run evolutionary strategy than of short-run economic tactics. It is a different order of decision. Expenditure of depleting and polluting geocapital (and the consequent evolution of technical change in that direction) is too different a thing from the capturing of nondepleting, nonpolluting solar income for a forced comparison in pecuniary units of discounted, abstract exchange value, not to be misleading. Obviously, it is always easier or "cheaper" to live off capital than off income—for as long as the capital lasts. It is not surprising that for most uses fossil fuels are cheaper than solar energy during an era of mineralogical bonanza. We hardly need have recourse to the arcane numerology of benefit–cost analysis to conclude that living on capital requires less conscious short-run sacrifice than living on income. The economics of living on permanent solar income, as all other forms of life do, differs from living mainly off terrestial capital as chess differs from checkers. The difference is deeper than the previous distinction between strategy and tactics would suggest. The very rules of the game are different though the board on which the two games are played looks the same. One game recognizes permanence and ecological discipline as rules restricting legitimate moves. The other game has no such rules. What is a good move in the checkers of geocapital consumption economics is usually not a good move in the chess of permanent, solar income economics. In checkers all pieces are comparable; in chess qualitative differences among pieces are essential. We simply must decide which game we want to play before we evaluate alternative tactics—that is, before moves based on pecuniary exchange-value calculations can serve our basic purposes, rather than pervert or obscure them. Market prices are relevant only to temporally and ecologically parochial decisions—decisions whose major consequences lie wholly within the human economy of exchangeable commodities and within the present generation. Market prices should not be used to decide the rates of flow of matter and energy across economy-ecosystem boundaries (depletion or pollution), or to decide the distribution of resources across generational boundaries. The first must be an ecological decision, the second an ethical decision. These decisions will, of course, influence market prices, but the whole point is that the ethical and ecological decisions should be viewed as *price-determining*, not *price-determined*.

The sophisticated tools of tactical decisionmaking become entirely sophistical if we try to use them to help us decide which game to play. To illustrate: solar energy would surely become cheaper than fossil fuel energy for most purposes if we decided to deplete the remaining fossil fuels at one-half the current annual rate. And what is to prevent us from gradually cutting the rate of depletion if we wish to live more off income and less off capital, to interfere less with the natural services of the biosphere, and to induce technical change to become both resource-saving and more dependent on renewables? Many would argue that such a substitution of solar for fossil fuels is uneconomic and, indeed, it would be if the calculations were made on the basis of prices valid under the old high rate of depletion. But we could just as well argue, on the basis of the new prices (those prices that would result if present depletion rates were cut by one-half), that going back to the old reliance on fossil fuels would be uneconomic. Both arguments are completely circular. The decision of whether to play permanent solar income economics or temporary geocapital consumption economics is *price-determining*, not *price-determined*. Should it turn out, as some argue, that the soft technologies are cheaper, even in terms of the price prevailing in the bonanza period of high geocapital consumption, we could then add an impressive a fortiori punch to the argument.[b] But we should not obscure the basic issue by admitting the opponents' grounds, even though it might allow a short cut to winning the argument, when those grounds are themselves misspecified. The basic decision for solar energy must be seen as a strategic, autonomous, price-determining decision, not a tactical, induced, price-determined decision.

But are resource prices really so arbitrary? What about competition and optimal intertemporal allocation, and all that? The value we assign to resources *in the ground*, before they enter the productive process, is largely arbitrary. Resources enter the market at their value in the ground, plus cost of extraction. Their value in the ground is arbitrarily set at zero in both capitalist and socialist economies. Thus, resources are valued mainly according to their cost of extraction. In a bonanza period, when extraction is easy and prices are minimal, the short-run supply is great and the price is low. The price may temporarily be driven even below the cost of extraction. Cheap energy from the petroleum bonanza has subsidized all other forms of mineral extraction. That the relative prices of many resources have shown a falling trend means that we have been enjoying the bonanza or waxing phase of the extraction cycle of many minerals. Such a trend does not mean that resources are becoming in any real sense

[b] Such a case has been made by Duane Chapman (1978), who showed that for space and water heating in California, solar energy is cheaper than fossil fuel or nuclear, once correction has been made for the implicit price subsidies and tax benefits conferred upon the fossil fuel and nuclear alternatives. But these arbitrarities are to be added to, not substituted for, those discussed above.

less scarce; it just means that for a while we were able to extract them faster than we could find new uses for them.

The essentially arbitrary nature of the price of resources in situ can be seen from the following reasoning. The price of resources in the ground is theoretically determined by supply and demand, which no one denies, but underlying the supply curve are cost curves, which require a definition of cost. Cost of production may be defined as actual historical costs paid, which in the case of resources in situ is zero, or as current or even future replacements costs, which for many resources would be very high, perhaps infinite. Further, cost can be defined as private or social. The temporal and the ecological horizons over which we try to trace out and account for the opportunity costs of our decisions must be arbitrarily cut off at some point. But cutting the chain just after the first link is too soon. On the demand side we face a similar problem. The market demand curve is the sum of the individual demand curves for the population. But what is the population of people whose demands are counted? Who is allowed to bid in the market? Obviously, future generations cannot bid in present markets. Suppose we arbitrarily expand demand to include our estimate of the needs of the next ten generations instead of arbitrarily limiting it to one? Again resource prices would be very high. Of course, poor people in the present are also excluded from the market, and arbitrary decisions about income redistribution also affect resource prices, as does the arbitrary geographical distribution of mineral deposits. The point is that, depending on essentially arbitrary conventional definitions of cost (historical versus replacement, private versus social), and on an arbitrary definition of the demanding population (present only, or some number of future generations), resource prices, theoretically determined by supply and demand, could range from zero to infinity.

In practice, the supply and demand curves of real markets select for the lowest price conventions—that is, for historical and private cost rather than replacement and social cost, and for consideration of the present and near present only. Therefore, the price of resources in situ is zero, at least until the generation in which exhaustion is imminent. The existence of market-determined rent (either differential rent or scarcity rent) does not really alter the picture. Differential rent depends on differential extraction costs, not on any concept of long-run replacement cost or on any consideration of future generations. It is a premium paid for greater accessibility and easier extraction, not a value imputed to resources in the ground. Scarcity rent in the sense of "user cost" (value of future use foregone) provides no escape from fundamental arbitrariness, since its magnitude depends on when in time the foregone future use is assumed to be felt. The shorter a community's time horizon, the lower will be user cost.

The claim that discounting future values to equivalent present value by means of the interest rate automatically counts the demands of the future and thereby achieves optimal intergenerational allocation is simply wrong. Present value calculations achieve optimal allocation as judged by the present generation. There is no reason to suppose that a future generation would consider such an allocation as acceptable, let alone optimal. In any case, "allocation" is not the proper word since different generations are different people, and division of resources among different people is distribution, not allocation. Distribution can be fair or unfair, but not "efficient." At most, the interest rate can aid the allocation of resources within the time span of a single generation. The fact that single generations form an overlapping chain into the distant future does not extend the interest rate's proper domain into the distant future. If the overlapping of the decisionmakers' lifetimes were the significant fact rather than the time horizon of the decisionmaker, we could reduce the time horizon to one year, or even one week and then argue that no one need think more than a week ahead, as long as the one-week periods formed an overlapping chain into the distant future. The absurdity is evident.

Suppose we were to change the rules of the game in favor of sustainability and calculate cost as nearly as possible on a replacement basis. Let renewable resources be exploited on a sustainable yield basis, and let nonrenewables be exploited at a rate such that the resulting price is at least as high as that of the nearest renewable substitute. For example, forests would be managed on a sustainable yield basis, with alcohol being made from trees. The extraction of petroleum would be limited to an amount such that the price per BTU-equivalent would be at least as high as that of wood alcohol. Resources in the ground would then have a positive value, probably a very high value, which could be captured by the government via a severance tax or sale of depletion quota rights. This would certainly be no more arbitrary than current prices. If one prefers the soft-energy future, then it would be a desirable policy for guiding the transition to a more sustainable economy, by means of altered relative prices.

If prices are to be good servants instead of bad masters, we must realize that the zero price of resources in situ is not a measure of relative scarcity objectively calculated by an impersonal market. Rather it is a conventional assumption on the basis of which the market determines all other prices and the resulting allocation and rate of use of resources. The price system should be viewed as a useful instrument to help us achieve efficiently whichever of the two energy futures we choose. It cannot make the choice for us. The choice between the hard and soft energy future, between a growth economy and a steady-state economy is price-determining, not price-determined.

The low-demand energy future is feasible. "Whatever exists is possible" (Boulding's First Law) is about as axiomatic as one can get. The low-demand scenario (one-half current U.S. per capita energy usage) *exists* today in Western Europe, and existed in the United States as recently as around 1960. Whether this lower total amount could eventually be supplied by soft technologies alone is less certain, but growing evidence suggests that it could be. Lovins' soft energy path envisions only a slight eventual reduction below current energy consumption, achieved very gradually, and at little or no sacrifice of end-use function, thanks to ample scope for efficiency improvement.

The common notion that the soft-technology, low-demand scenario is "far out" or merely a hypothetical polar case is due to inability to recognize the obvious. It is the high-demand, hard-technology future that has never before existed and is completely hypothetical. Yet self-styled "realists" all treat the hypothetical high-energy scenario as if it were empirically verified, and treat the empirically verified, already experienced low-energy case as if it were the flimsiest conjecture!

Of course, the transition to a soft-energy future should be gradual, and of course, there is room for compromise. But the basic question must be put in a clear-cut way, and a definite either/or decision be made on the *direction*, even though the *rate* at which we move in that direction is subject to compromise and economic constraint. Many people do not like to face up to this basic choice because it is not a question of rationality of means, but of sanity of ends. Taking a position requires more than counting; it imposes moral self-definition and responsibility.

Dickinson's *Letters of John Chinaman* describes the condition very well, as seen from the viewpoint of an Oriental spectator looking at England, early in this century:

> Your outer as well as your inner man is dead; you are blind and deaf. Ratiocination has taken the place of perception; and your whole life is an infinite syllogism from premises you have not examined to conclusions

you have not anticipated or willed. Everywhere means, nowhere an end. Society is a huge engine, and that engine itself, out of gear. Such is the picture your civilization presents to my imagination (Quoted in Pigou 1932: 13).

We are still in the same fix: everywhere means, nowhere an end; everywhere market prices, nowhere moral values. Unless we face up to the four questions of purpose, and reason our way toward some minimal moral consensus, then it does not really matter what our energy policy is.

WHENCE MORAL CONSENSUS?

But where is this moral consensus to come from? Not from a spineless relativism or from the hallucinatory psychic epiphenomena of complex mechanisms. Let us state it directly in the strongest terms. Ultimately, the possibility of moral consensus presupposes a dogmatic belief in objective value. If values are subjective, or thought to be merely cultural artifacts, then there is nothing objective to which appeal can be made, or around which a consensus might be formed. Consensus based upon what everyone recognizes to be a convenient cultural myth (like belief in Santa Claus) would not bear much stress. Only real objective values can command consensus in a sophisticated, self-analytical society. We have no guarantee that objective value can be clarified, nor that once clarified it would be accorded the consensus that it merits. But without faith in the existence of an objective hierarchy of value and in our ability, at least vaguely, to perceive it, we must resign ourselves to being driven by the force of technological determinism into an unchosen and, perhaps, unbearable future. On what other grounds is technical determinism to be resisted? Faith in the existence of objective value is to public policy as faith in an orderly universe is to physical research.

In the words of C.S. Lewis, "A dogmatic belief in objective value is necessary to the very idea of a rule which is not tyranny, or an obedience which is not slavery" (1973: 329). The same insight underlies Edmund Burke's famous dictum that "society cannot exist unless a controlling power upon will and appetite be placed somewhere, and the less of it there is within, the more there must be without" (1865: IV, 51–52). Control from within can only result from obedience to objective value. If interior restraints on will and appetite diminish, then exterior restraints, coercive police powers, or Malthusian positive checks must increase. In Burke's words, "men of intemperate minds cannot be free. Their passions forge their fetters."

The major reason for pessimism about the course of human affairs is that the very term, "*dogmatic* belief in *objective* value," automatically shuts the minds of most modern intellectuals.[c]

G.K. Chesterton informs us that:

To be dogmatic and to be egoistic are not only not the same thing, they are opposite things. Suppose, for instance, that a vague skeptic eventually joins the Catholic Church. In that act, he has at the same moment become less egoistic and more dogmatic. The dogmatist is by the nature of the case, not egotistical, because he believes that there is some solid, obvious and objective truth outside him which he has perceived, and which he invites all men to perceive. The egoist is, in the majority of cases, not dogmatic because he has no need to isolate one of his notions as being related to the truth; all his notions are equally interesting, because they are related to him. The true egoist is as much interested in his own errors as in his own truth; the dogmatist is interested only in the truth, and only in the truth because it is true. At the most, the dogmatist believes that he is in the truth, but the egoist believes that the truth, if there is such a thing, is in him.

A related clarification was made by E.F. Schumacher:

The result of the lopsided development of the last three hundred years, is that western man has become rich in means and poor in ends. The hierarchy of his knowledge has been decapitated: His will is paralyzed because he has lost any grounds on which to base a hierarchy of values. What are his highest values?

A man's highest values are reached when he claims that something is good in itself, requiring no justification in terms of any higher good. Modern society prides itself on its pluralism which means that a large number of things are admissible as "good in themselves," as ends rather than as means to an end. They are all of equal rank, all to be accorded *first priority*. If something that requires no justification may be called an "absolute," the modern world, which *claims* that everything is relative, does, in fact, worship a very large number of "absolutes" . . . Not only power and wealth are treated as goods in themselves—provided they are mine and not someone else's—but also knowledge for its own sake, speed of movement, size of market, rapidity of change, quantity of education, number of hospitals, etc., etc. In truth, none of these sacred cows is a genuine end; they are all means parading as ends. (1977: 58)

[c]My pessimism on this score was reconfirmed by the resistance to the notion of objective value that was forthcoming from various quarters at this symposium. One critic followed his denial of objective value to its logical conclusion and denied that there is any such thing as "the public interest." If true, that would make nonsense of the whole conference on "National Energy Issues— How Do We Decide?" since we could not possibly decide by appeal to a nonexistent public interest, and private interests can fight things out on their own without academic symposia.

Science and technology, with their analytic-empirical mode of thinking, have led many into a kind of scientism which seeks to debunk all knowledge that does not have an analytic-empirical basis. Knowledge about ends—about objective value and right purpose—derives from an "illicit" source and is considered "forbidden knowledge" by the priests of the scientistic inquisition. Unless this error is recognized, unless we come around to a "dogmatic belief in objective value," or what Boris Pasternak called "the irresistible power of unarmed truth" then it makes no sense to concern ourselves with economics. Why strain at gnats of marginal inefficiency in the allocation of means to serve ends, while swallowing camels of total incoherence in the ordering of those ends? Indeed, if our ends are perversely ordered, then it is better that we should be *inefficient* in allocating means to their service.

TOWARD A PRINCIPLE OF RIGHT ACTION

It is one thing to insist on the logical necessity of a dogmatic belief in objective value as a basis for resisting technical determinism, but it is something else to have clear and certain knowledge of what objective value is. We must be open-minded regarding differing understandings of the nature of objective value and corresponding principles of right action. But we must make an effort to state the general principles that should guide our decisions and apply them to energy.

Probably the rule of right action most accepted in practice is Jeremy Bentham's "greatest good for the greatest number." Economists have avoided the difficult problem of defining "good" by substituting the word "goods," in the sense of commodities. The principle thus became "the greatest per capita product for the greatest number." More products per capita and more people to enjoy those products, lead, in this view, to the greater social good. Our commitment to growth is no doubt based, in considerable degree, on this principle which implies that right action is that which leads to more good for more people.

But there are two problems with "the greatest per capita product for the greatest number." First, as others have pointed out, the dictum contains one too many "greatests." It is not possible to maximize more than one variable. It is clear that numbers of people could be increased by lowering per capita product, and that per capita product could be increased by lowering numbers, since resources taken from one goal can be devoted to the other. Second, it makes a big difference whether "the greatest number" refers to those simultaneously alive, or to the greatest number ever to live *over time*.

To resolve the first of these difficulties we must maximize one variable only, and treat some chosen level of the other as a constraint on the maximization. For one of the "greatests" we must substitute "sufficient." There are two possible substitutions: the greatest per capita product for a sufficient number; or a sufficient per capita product for the greatest number. Which is the better principle? I suggest that we adopt the latter, and that "greatest number" be understood as the greatest number ever to live over time, which takes care of the second problem. The revised principle thus becomes "sufficient per capita product for the greatest number over time."

It is hard to find any objection to maximizing the number of people who will ever live at a material level sufficient for a good life. However, this certainly does *not* mean maximizing the number alive at any one time. On the contrary, it means the avoidance of any destruction of the earth's capacity to support life, a destruction that results from overloading the life-support system by having too many people, especially high-consuming people, alive at once. The opportunity cost of those extra lives in the present is fewer people alive in all subsequent time periods and, consequently, a reduction in total lives ever to be lived at the sufficient level. Increasing per capita product beyond the sufficient level (extravagant luxury) may also overburden life-support systems and have the same long-run life-reducing effect as excess population.

Maximizing number while "satisficing" per capita product does not imply that quantity of life is a higher value than quality. It does assume that beyond some level of sufficiency, further increase in per capita goods does not increase quality of life, and in fact may well diminish it. But sufficiency is the first consideration. To put it more concretely, the basic needs of all present people take priority over future numbers, but the existence of more future people takes priority over the trivial wants of the present. The impact of this revised utilitarian rule is to maximize life, or what is the same thing, to economize the long-run capacity of the earth to support life at a sufficient level of individual wealth. One is reminded of Thoreau's statement, "The cost of a thing is the amount of what I will call life, which is required to be exchanged for it immediately, or in the long run." The sufficient level may be thought of as a range of limited inequality rather than a single specific per capita income applicable to everyone. Some inequality is necessary for fairness.

If one accepts this revised utilitarian principle, how does it apply to the test case of plutonium? If a sufficient per capita product for a sustainable population can be achieved without plutonium, then why risk significant destruction of life-support capacity by moving to a plutonium economy? Advocates of the plutonium economy might argue that, if everything works as planned, the additional energy source could increase the number of lives ever to be lived at the sufficient level, and therefore is to be desired. But then one must ask, what if those same resources were devoted to improving our means for tapping solar energy? Might not the latter increase life-support capacity even more over the long run, since the sun will outlast uranium supplies by a very long time, and since the consequences of error and malice are very small with solar energy and very large with plutonium?

This modified utilitarian principle certainly offers no magic philosopher's stone for making difficult choices easy. But it does seem superior to the old Benthamite rule (implicit in much of our decisionmaking) in that it draws our attention to the concept of sufficiency, and extends our time horizon. It forces us to face the question of purpose: sufficiency *for what?* Needed for what? It will be very difficult to define sufficiency and build the concept into economic theory and practice. But I think it will prove far more difficult to continue to operate on the principle that there is no such thing as enough.

REFERENCES

Burke, Edmund. 1865. "Letter to a Member of the National Assembly." In *The Works of the Right Honorable Edmund Burke.* 12 vols. Boston: Little, Brown.

Chapman, Duane. 1978. "An Analysis of Federal Incentives Used to Stimulate Energy Production." Battelle Pacific Northwest Laboratories, PNL−2410. March.

Chesterton, G.K. n.d. "Introduction." In *Poems by John Ruskin.* London: Rutledge.

Daly, H. 1977. *Steady State Economics.* San Francisco: W.H. Freeman.

Ford Foundation, Energy Policy Project. 1974. *A Time To Choose.* Cambridge: Ballinger Publishing Company.

Galbraith, J.K. 1958. "How Much Should a Country Consume?" In *Perspectives on Conservation,* edited by Henry Jarrett. Baltimore: Johns Hopkins Press.

Georgescu-Roegen, N. 1971. *The Entropy Law and the Economic Process.* Cambridge, Mass.: Harvard University Press.

Lewis, C.S. 1973. "The Abolition of Man." In H.E. Daly, ed., *Toward a Steady-State Economy.* San Francisco: W.H. Freeman and Co.

Lovins, Amory. 1977. *Soft Energy Paths.* Cambridge: Ballinger Publishing Company.

Pigou, A.C. 1932. *The Economics of Welfare.* 4th ed. London: Macmillan.

Schumacher, E.F. 1977. *A Guide for the Perplexed.* New York: Harper and Row.

Soddy, Frederick. 1922. *Cartesian Economics.* London: Henderson.

A Discussion of the Technical Implications of the Growth Scenario Presented by René H. Malès and the No-Growth Scenario Presented by Herman E. Daly

Robert V. Laney

We are greatly indebted to René Malès and Herman Daly for their clear, interesting presentations. I found them both most stimulating. I hope I will be forgiven if, in order to address the purpose of this symposium, I now dwell on their shortcomings. Fortuitously, perhaps, Drs. Malès and Daly have, at the very beginning of this symposium, portrayed our dilemma, at least as I see it. Their two papers faithfully adopt the same polarized viewpoints and repeat the same exhortations which have thus far confused and frustrated a public which ought to be making up its mind about nuclear power. Like signals sent on different frequencies, their messages mingle, but rarely touch. Each speaker presents well-constructed and thoughtful discussions, addressing the topic, "Decision Criteria for Needs for Nuclear Fuel." Yet each adopts a perspective based on his own, sometimes unstated, assumptions and value judgments and proceeds to his own predictable conclusion. In their chosen vernaculars, they say much to us, but little to each other. After consideration, we realize that we have witnessed a demonstration of the intellectual insularity which characterizes the nuclear debate.

Let us examine the views expressed by the speakers and try to understand why they fail to offer us a satisfactory basis for comparing alternatives and fail possibly even to help us select the more persuasive one. Dr. Daly provides a useful framework on which to structure this examination. He points out: "We must guard against the twin errors of *wishful thinking* (assuming that whatever is desirable must somehow, someday be possible), and *technical determinism* (assuming that whatever is possible must somehow, someday be desirable)." As we hereafter use the terms "desirable" and "possible," we should keep in mind that assessing desirability requires value judgments generally not quantifiable, whereas estimating what is possible involves technical and economic evaluations usually addressed in quantitative terms.

After providing us with the convenient "desirable/possible" structure, Dr. Daly does not seriously take up the challenge of gauging what is possible. Instead he assumes that the "low demand, soft technology" path, accompanied by stabilized or reduced population, will be possible if only we want it enough. There is, in my view, a serious omission of both technical and economic analysis to support his proposed reliance on soft energy technology; he provides only footnote references to the Ford Foundation's *A Time To Choose*, and to Amory Lovins' *Soft Energy Paths*. Neither of these references presents the substantive analysis which would be needed to justify their or Dr. Daly's conclusions about what is possible.

Even when expressing his preference for the soft path, Dr. Daly does not compare its desirability with other possible futures. He argues, following Lovins, that the two paths, centralized or decentralized, hard or soft, cannot coexist. We must choose between them. Technically, this position leaves out the infinitely more likely middle ground. After all, we are today using both centralized and decentralized forms of energy; both uses are growing, and both are certain to continue to grow. True, there are technical and institutional difficulties with linking decentralized solar installations with centralized energy production, especially in the generation of electricity. But there is much to suggest that these linkages will be developed successfully and little to suggest that these must be mutually exclusive options. By focusing on the energy patterns of a distant, idealized world and largely neglecting the vexing problems of transition, Dr. Daly's presentation fails to confront or provide guidance for the difficult choices which must be made now, next year, and in the remaining years of this century.

René Malès offers a quite different perspective. He apparently assumes the inevitability or the desirability of continuing centralized energy growth and devotes most of his discussion to presenting a methodology for deciding how fast centralized energy should grow in the face of uncertainties in energy demand, economic growth rate, and conservation effectiveness. In Dr. Daly's terms, Dr. Malès might be thought guilty of "technical determinism," that is, of assuming that since centralized energy growth is possible, it must be desirable. By deriving planning targets from historical data, by assuming an unbroken continuation of present patterns of energy use, by failing to examine whether there may be long-term disincentives for this course of growth, and finally, by concentrating largely on the remaining years of the century, Dr. Malès presents something very much like the inverse of the Daly viewpoint. Where Dr. Daly projects us somehow willy-nilly into a radically changed, although undetailed, world of the distant future, Dr. Malès leaves that distant future unexamined.

I must note, however, the valuable contributions Dr. Malès makes in providing methods for introducing probabilities into planning models, for analyzing the nonsymmetry of wrong decisions (for we may be sure of wrong decisions), and for introducing (to me at least) the most useful discussion of energy and capital as substitutes or complements.

The two presentations thus embody our dilemma: today's energy decisions are determined largely by existing imperatives of population, jobs, and human aspirations; yet these decisions also foreshadow, perhaps irretrievably, the long-range future. Nuclear power and the breeder reactor, in special ways, present this prospect of irreversibility. Unfortunately, it seems to me, neither speaker deals simultaneously with both the energy choices of the next 25 years and the long-range implications of these choices. In addition, since they have focused their thoughts on different time periods and have assumed different answers to "what is desirable," their two papers, when taken together, do not help us very much with this central quandary.

Since our objective in this symposium is "to define a program of inquiry leading to a better understanding of the methodological and philosophical conflicts with which energy policymaking is so sorely beset," I will suggest certain modifications in the perspective of the preceding presentations that might place them along the same locus of thought and thus enable a concerned layman to understand better where the speakers differ, what choices are before us, and what is the weight of evidence supporting the choices.

For those who, like Dr. Daly, espouse low growth and soft technology, I ask that you address the problem of effecting a safe transition from today's high consumption, centralized-generation society, to that distant renewable energy world which you visualize. However desirable that future world may appear and however successfully you may persuade the rest of us of its desirability, we know that the pathway to it must lead from where we are and what we are today. You will encounter less doubt about the desirability of your vision of the future than about its attainability.

You suggest that we must level, or possibly reduce the world population. Most of us could agree with this. How do you think we might go about achieving such a leveling? What are we to do in the meantime about the *fact* of continued population growth? Also, you suggest that by arbitrarily reducing the depletion rate of fossil fuel "solar energy would surely become cheaper than fossil fuel energy for most purposes." Since this would presumably occur as a result of a substantial increase in the cost of all energy, we should be told how and when and at what rate this might take place, to what new energy levels the economy must adjust, and what might be the consequences for life-styles. If there are implicit assumptions of new solar technological advances, we should understand what these assumptions are and on what they are based.

Finally, there should be more adequate explanation of your insistence on a "definite either/or decision" on the soft-technology, low-demand scenario. History teaches us that major transitions of the kind you envision will proceed slowly, continuing the use of tried and reliable methods while developing new ones. The suggestion that we loose our grip with one hand before we have a firm grasp with the other is not an acceptable proposition and raises questions about the credibility of those who propose it. Again, we need a more explicit consideration and analysis of the transitional period.

To Dr. Malès and others who accept, without evident question, the desirability of continued growth of centralized energy systems, we ask for greater consideration of the long-term effects of the centralized technology and the continued economic growth which this implies. Are we, by present decisions in favor of traditional growth patterns, living out a self-fulfilling prophecy? Do today's decisions commit us irreversibly to long-term consequences which may be found unacceptable by future generations? Can we freely make midcourse corrections whether for environmental reasons or to utilize new, better technology, or are we establishing irrevocable "trends to destiny"? We must question more searchingly the deterministic quality of present growth decisions. We need a more adequate vision of where the mixed hard and soft technology course leads.

Those who share either of these alternative viewpoints should provide more substance. The soft-energy spokesmen should give us suggestions for making current energy-related decisions, showing how, in a continuous time frame, these may lead towards the more distant future you envision. From the centralized-production, continued-growth advocates, we ask a more detailed consideration of the fission-energy-based world of the twenty-first century, showing how its special problems will be resolved.

Peter Drucker writes, "Long-range planning does not deal with future decisions, but the future of present decisions."

We ask René Malès, "What is the *future* of the energy decisions you propose?"

We ask Herman Daly, "What *present* energy decisions do you recommend?"

Rebuttals by Speakers
Herman Daly and René Malès
to Discussant Robert Laney

Herman Daly: "What is the safe transition to a renewable energy future?" I think that an either/or discontinuity in a decision does not imply a historical discontinuity of social evolution. I do not think you can argue that there is a discontinuous social evolution implied by saying we go either the soft or the hard energy path. The discontinuous decision I am talking about is a question of principle, a question of direction that people have to decide. I accept whole-heartedly the notion that we are where we are, and we have to take *that* as a starting point, rather than an imaginary blank slate. Surely in a technical sense, as Amory Lovins has emphasized in his writings, there is no incompatibility between the soft and hard paths; the two obviously coexist right now and will have to coexist in the future. The incompatibility is one of political commitment to one direction or to the other; it is hard to ride two horses going in different directions. You have to let one take the lead, at least marginally, and push the other one into the background.

As for the possibility of a low-demand, stable system, again I think that this is a possibility for which there is evidence over history. What has really not ever been experienced historically is the high demand, totally centralized, all-electric sort of economies that are envisioned as the alternative. There is no empirical case to point to of a plutonium economy, yet there are many traditional societies that have lived on low energy budgets in a renewable way. There is Western Europe, which, at least in terms of energy consumption, as Dr. Malès indicated, is considerably more efficient than we are. In that sense, I recognize the importance of his criticism, but I think it can be answered, and it is rather harder for the high-growth position to answer that question, than for the low-growth. After all, whatever exists is possible, and the low-demand scenario does exist. Advocates of the plutonium economy cannot make that statement. It seems to be a little bit asymmetrical there.

"To what energy price level must we adjust?" Well, I really don't know the answer to that, but I have one sort of general principle: for the sake of sustainability, we should gradually shift towards renewable sources. I don't advocate any kind of drastic reduction or anything that will cause violent changes. I do believe in a gradual kind of shift. I think we should eventually adopt a sustainable yield principle for our renewable resource base. Whatever price results from using resources on a sustainable yield basis, that is the price towards which we eventually have to move. We might as well use that price as a criterion for prices of nonrenewables as well; the price of a nonrenewable resource should be at least as high as that of the closest renewable substitute, when that substitute is exploited on a maximum sustainable yield basis.

The transition is important, but I don't think that we can talk about a transition until we have some agreement on what it is a transition *to*. We know where we are, but there is enormous disagreement about where we want to go. Granted that the transition is extremely important, a first step in dealing with it is to clarify further where we want to go.

René Malès: I'm afraid I did not make myself clear. I did not mean to advocate high growth or advocate hard technology for the sake of hard technologies. The position I was trying to express was that this was a feasible, possible scenario of the future. Further, there is a substantial cost to society of not making it a feasible technical future, if in fact, it is desirable. To some extent, the questions of plutonium and breeder availability are the questions of whether such a society would be technically feasible.

Dr. Laney made several other important points. The first is the question: "Is it endurable?" The second one is, "Is it a desirable future or an acceptable future?" Both are questions that ought to be asked. But I think these are separate from the question I was answering as to whether a high-growth future was a feasible and possible one.

Chapter 3

Deciding Energy Policy: Some Socioeconomic Implications

Lester B. Lave

Symposia being what they are, I will err on the side of being provocative and outrageous, rather than being agreeable and dull. Drs. Malès and Daly have presented (in Chapters 1 and 2) many targets of opportunity for being outrageous, but I cannot resist opening with a shot at Alvin Weinberg, our session chairman. Dr. Weinberg asked us to engage in clear, logical discourse and avoid the sort of emotional outbursts that have clouded the nuclear debate. Certainly I agree that our intellectual discourse ought to be calm and closely reasoned. But this perspective does not recognize the nature of democracy and the way that society makes decisions. Democracies are characterized by almost infinite inertia. To change a social policy or to get a large number of people to think about it, one must get their attention. Unfortunately, that seems to require overstatement and many noncalm, nonintellectual methods. I find myself glad (at least ex post) to be compelled to think about some issue that I had thought myself too busy to consider and glad that society is taking seriously some question that otherwise would never receive wide consideration. I would hope that we would all be calm and reasonable within this meeting, even though some of us might not choose to be calm and reasonable outside it.

MALÈS' ANALYSIS

Let me begin with some comments on René Malès' analysis, in which a central role is performed by the social loss function, that is, society's loss from having too much or too little energy. I am not convinced either that long-run shortages of electricity are much more costly to society than long-run over-capacity, or that either is terribly expensive. Furthermore, the estimated cost of shortage makes no sense. I must agree with an earlier comment that it is inappropriate to extrapolate the losses associated with an outage to the losses that would be expected from having less energy than is desired in the future. The basic reason stems from a principle that economics borrowed from physics. The "Le Chatelier" Principle states that in maximizing a function subject to constraints, relaxing one of the constraints will mean that the value of the objective function cannot fall and in general will rise. In this case, the constraint is the time given to react to getting less power than is desired. For an outage, one finds the loss to be immediate and total. Surely such costs are much higher than those resulting from knowing now that in five years there will be 20 percent less electricity than will be desired. Unless we are suicidal, we will find a way to keep the system from failing and find an allocation plan via price increases or rationing, for the electricity that is available. Furthermore, steps will be taken to increase the efficiency with which electricity is used. Under a circumstance such as learning now that there will be a shortfall in electricity generation in five years, I am confident the loss would be a great deal smaller than the cost of an outage.

There is not so much of an asymmetry between the costs of over and under capacity as Dr. Malès states. According to Say's Law, "Supply creates its own demand." Although Say's Law is bad economics, it is good psychology: having a large investment induces people to work hard to see that the investment is used. If we have built an unnecessary power plant, I suspect we would see Herculean efforts to find ways to use all that power. I cringe at the notion that we would build a number of $1 billion (1 gigawatt) plants "just in case." That is a great deal of capital, and we can ill afford to be unwise with such large sums.

My next point applies not only to Dr. Malès' analysis, but quite generally, even to some of my own work. "A quad isn't a quad . . ." Gertrude Stein notwithstanding. I confess to having collaborated in this exercise of looking to the future and adding a quad of energy from coal to a quad of energy from electricity. That is simply nonsense. The quads cost different amounts to produce and are worth very different amounts in consumption. It is misleading to add together quads without weighting them by their incremental use value (price in a free market).

In particular, the graph showing the ratio of energy to GNP over time does not account for the quality of the energy. In switching from wood to coal, to oil and natural gas, the quads have become much more valuable. I wonder if the variation in the energy GNP ratio over time would be quite so large if one adjusted the quality of the energy.

Having helped Dr. Malès' point a bit, I now want to quarrel with him. You will remember that "There are lies, there are damn lies, and there are statistics." After the forecast energy demand, backed up by "good econometric analysis," I am tempted to amend the statement: "There are lies, damn lies, and statistics, but in a class by themselves are econometrics."

I simply see no way that U.S. energy use will grow 5.3 percent per year by 2000. A theme that I will elaborate in a moment is the increased cost of energy and the recognition of the social costs of producing energy. Even if one were highly optimistic about the economy (which I am not), one would not forecast such high growth rates of energy use for these reasons.

Dr. Malès tended to be a bit slippery in mixing up demand and need. If by "need" one meant the amount of energy needed to sustain life, that amount is so small that it is irrelevant to our considerations. Even if one meant the amount of energy needed to sustain per capita income, that amount is small enough so that it could be produced without great difficulty. Instead, what I want is "need," whereas what you want is merely demand.

By and large, the amount of energy produced by the United States economy over the next several decades will result from the interaction of the demand for energy with the supply of energy. Production will be above anything one might legitimately call need, and so this concept is simply irrelevant. I suggest we banish that word and focus instead on the amount of energy we desire, in view of its cost and externalities.

DALY'S ANALYSIS

As I reach into my pail of mud for the next shot, I feel that it would be ungracious of me to neglect Dr. Daly, who appears to be attacking a straw man of exponential growth. Whatever its faults, Meadows, Meadows, and Behrens (1972) convinced us that exponential growth is incompatible with a finite world. Surely the interesting issue is, "When enough is enough." Should we have stopped growing in 1600, or can we afford to grow until 2050?

In reading Dr. Daly's paper, I found myself wondering what a steady state is. From an anthropological viewpoint, one could have a steady state with a few dozen stone-age hunter-gatherers, living in an area of hundreds or thousands of square miles. Alternatively, one could take a stone-age agricultural society with population densities of a few hundred people per square mile. Or, one could leap to a society using twentieth-century technology and living in a steady state along the lines Dr. Daly suggests. My point is that the improvements in technology not only increase population density, but increase the standard of living.

During the last two centuries, when we have made the most profligate use of our natural resources, we have made the greatest technological progress. I do not think the two were unrelated. If you will agree with that, I wonder what our great grandchildren would prefer us to do? Would they prefer that we give up technological progress and go into a steady state now, or would they prefer additional technological change coupled with using up most of the remaining oil and natural gas? The answer is not obvious, or if it is, technological progress would seem to win.

Dr. Daly emphasized that we have become a culture that examines means rather than ends. As a social scientist, I would agree with that observation and its corollary: that determining social policy is much more difficult under such circumstances. If everyone agrees that the goal is construction of a pyramid, or following a particular religion, then social policy is easily determined. The last two centuries have destroyed any semblance of a generally accepted set of social goals. Democracy, civil liberties, and free market capitalism pushed out the old ideals. But even the worship of wealth and success died with Social Darwinism at the beginning of this century.

There is a sort of social Gresham's Law where affluence and democracy drive out previous goals, but rapidly decay into footloose hedonism. "Give us this day a sunny beach and a six-pack of beer." While many people want something beyond hedonism, there is no agreement on what that is; we have no agreement on goals.

What we are left with is an elaborate set of mechanisms for determining due process. Having failed to agree on our energy policy, we spend our time producing unread environmental and inflationary impact statements. The mountains of paper camouflage the fact that we have not agreed on our goals. We have succeeded in elaborating due process to the point where it is virtually impossible to do anything. But perhaps that is the point. Having failed to agree on goals, we seem to be busily attempting to prevent anything, good or bad, from being done.

In one sense Schumacher, Lovins, and Daly are attempting to lead us to considerations of goals. They are taking us away from environmental impact statements and asking us about our ultimate goals;

I admire their courage, but I cannot believe that they will be successful in convincing most people. Western society in general, and American society in particular, is so diverse that I cannot believe that reasoned discourse can possibly get us to agree on goals.

Indeed, as I understand the Schumacher-Lovins-Daly argument, its advocates are attempting to gain adherents not by stressing ends, but rather by attempting to define attractive means. For example, a steady state is a means, not an end. There are a vast number of different societies which are compatible with a steady-state use of resources. Rather than select one such society, Dr. Daly broadens his appeal by concentrating on the more attractive process goal of a steady state. Thus, each listener is free to imagine what goals the steady-state society would satisfy. I am afraid that Dr. Daly illustrates his own argument only too well; one cannot get agreement of goals, only on processes. Unless we give up democracy and are willing to march blindly to some drummer, then I am afraid we will have to make social policy without having agreed-upon goals.

HOW SHOULD WE DECIDE?

While it is fun to fire darts at Drs. Malès and Daly, my conscience reminds me of the title of this symposium. I feel compelled to address a few remarks to that subject. Society has indicated the need for some soul searching regarding the amount of energy use we desire and the resources and technology that will be used to provide it. I must disagree with Dr. Daly and assert that the framework he rejected is a fruitful one for addressing the issues.

It is not that this framework has failed us, but rather that it has only recently begun to be used. Until recently, we made no attempt to consider safety, health, and ecology. We now know a great deal about the social costs, both short and long term, of using various resources and technologies to produce energy. Those costs are much higher than had been suspected. The costs stem not merely from plutonium or other nuclear technologies, but from all energy technologies. Even such benign fuels as oil and natural gas have such important social costs as oil spills, air pollution, and accidents. By failing to recognize these social costs up to now, we were in effect subsidizing energy.

We have learned a great deal, and it seems unlikely that we will continue to make the same mistakes by subsidizing energy in the future. Rather, society will find a way to increase the costs of energy production to reflect the externalities. Nor should anyone have the impression that the finger is being pointed at nuclear technologies.

All technologies, from nuclear to solar, have important externalities. All will have their costs increased by attempts to mitigate the untoward effects. Nor is it obvious that the technologies such as solar, that are widely regarded as benign, will turn out to be cheaper after controlling for untoward effects than coal or nuclear technologies. The prime lesson is that energy will be a great deal more expensive in the future, because of correction of externalities, because of greater costs of extraction, and so forth.

Every energy resource and technology has important externalities. For example, biomass requires large quantities of vegetation to be grown for burning or fermentation. This requires extensive cultivation, irrigation, and sweeping ecological changes. Photovoltaic energy requires papering over large expanses of land. Energy use will surely grow more slowly in the future.

FORECASTING ENERGY DEMAND

What will be the demand for energy in the future? The three factors most important in shaping future energy demand are population growth, technological change, and attitudes toward work and consumption. As a social scientist, I have to confess that we know almost nothing about the factors determining any of these three.

The fertility rate in the West has been falling steadily. In the United States the fertility rate is below the lowest level reached in the 1930s and is still falling. Even if the rate were to stop falling, we would still not be reproducing ourselves and would have to look forward to a smaller population. Energy demand in the twenty-first century will depend importantly on whether the fertility rate continues to fall, levels off, or rises. There are no good ways to forecast what will happen, unfortunately. Periodically, some demographer tries to be a sage by predicting that the fertility rate is about to rise; so far, each sage has gotten tea leaves on his shirt.

Over the last century, technological change in the United States has averaged about 2 percent per year. If that continues, the demand for energy will be much higher than if it slows. While a good deal of work has been done to find the sources of technological change, almost nothing is known about what might be done to speed it up or to forecast the future. Some of my colleagues are highly optimistic, while others are much more pessimistic.

Attitudes toward work and consumption will decide future production. Will the length of the work week decrease? Will people join the labor force in greater numbers and for longer time periods? Again, because little is known, forecasting the future is little more than guessing.

For the next decade or two, it seems a safe bet that GNP and energy use will grow more slowly than over the first seven-tenths of this century. An even safer bet is slower growth over the next millennium. But the first half of the next century appears highly uncertain.

SEPARATING ENERGY POLICY FROM SOCIAL REFORM

Perhaps I can shake off a bit of my gloom in making a last point. Since the early 1970s energy policy has been taken to encompass virtually all questions about long-term goals. To some people, energy policy means that it is immoral to drive a gas-guzzling automobile. To others, energy policy means recycling to conserve resources. To others, energy policy means living a life which will provide minimal disruption to other living organisms. If we insist on settling all such issues before deciding whether to build a power generating facility, then we will build nothing. We have created paralysis in the name of energy policy.

This is silly for many reasons. Paralysis means that oil imports grow each year and that we use more of our most precious resources. Nor is there any necessity for actually resolving many of these issues. Our pluralistic society can easily accommodate many approaches, and it seems doubtful whether these issues all will be resolved or will need to be resolved.

Some of the issues must be confronted, and within five to ten years we can hope for some workable compromises. It would be folly to put off all action until these are resolved. We need to take steps now.

I would suggest that we divide energy issues into three categories: (1) those deep questions that need not be resolved, issues that ought to be the subject of undergraduate debate; (2) those issues that we might hope to resolve within a decade, which should be the subject of general study and debate; (3) those issues that are not really controversial: we know that we want to decrease oil imports and build facilities that will curtail emissions. I suggest that there are enough issues in this third category to provide a basis for taking action now to resolve many of the pressing social problems. Let's get on with it!

REFERENCE

Meadows, D.H.; D.L. Meadows; and W.W. Behrens, III. 1972. *Limits to Growth.* New York: Universe Books.

Chapter 4

Economic Growth and National Energy Policy: Some Political Facts of Life

Kenneth A. Shepsle

The energy debates of recent years appear to have attracted a number of split personalities. Rather than Jekyll and Hyde, however, the personalities are, in varying proportions, a combination of Chicken Little and Dr. Pangloss. "The sky is falling! The sky is falling!" is the initial response to (choose one of the following): (1) projected energy demand given current consumption patterns, or (2) expected energy supply in regulated markets, or (3) anticipated environmental degradation and feared health consequences either with the current dependence on fossil fuels or with nuclear energy, or (4) prospective nuclear proliferation and terrorism in a politically precarious world with nuclear-fueled energy. Life looks "nasty, brutish, and short," as the seventeenth-century political philosopher Thomas Hobbes said. A solution, however, exists in this melodrama in Dr. Pangloss' public policy laboratory. Everything will be copacetic if we simply (again, choose one or more of the following):

1. Dampen the demand for energy by taxing high-energy use; subsidizing energy-efficient schemes; reducing population pressures for more energy; and changing current attitudes tastes, and preferences that link the good life with energy consumption;
2. Increase the supply of energy by deregulating prices in various energy markets; relaxing certain environmental standards; and providing tax incentives, subsidies, and other forms of assistance to encourage more exploration, more expensive extraction, and more research and development of energy alternatives;

3. Halt or reduce environmental degradation and adverse health effects by the heavy taxing, regulating, or outright prohibiting of irreversible extraction techniques, on the one hand; and the location, rate of utilization of, and disposal of by-products from fuels harmful to health and environment, on the other;
4. Minimize the likelihood of nuclear accident by "safeguards in depth" and the likelihood of proliferation by careful monitoring of our policies toward exporting technologies capable of being put to nonpeaceful uses.

Of course, both Chicken Little's list of fears and Dr. Pangloss' collection of remedies could be expanded. Everybody probably has a little bit of Chicken Little and of Dr. Pangloss in him. One of my concerns as a political scientist is what it is about our political process that encourages the Chicken Little and Dr. Pangloss parts of us to dominate political decisionmaking. This is a point to which I shall return briefly at the conclusion of my paper.

I wish to focus on the kinds of trade-offs implied by the choice of an energy policy and the way these trade-offs are formulated by our political processes and institutions. I address the electoral connection, the institutional connection and the public interest connection.

THE ELECTORAL CONNECTION

No national policy, and, a fortiori, no national energy policy, is very far removed from the two-year and four-year electoral cycles. The reason is fairly straightforward. Our congressmen, senators, governors, and presidents are political animals who, for a variety of reasons, seek to attain and retain office. To the extent that it is within their control (and often it is not), they time their activities to peak in even-numbered years, preferably as close to, but not beyond, the first Tuesday after the first Monday in November. Additionally, they seek to generate benefits for *their* constituents in even-numbered years, either deferring costs indefinitely (in effect, giving their own constituents a free lunch), or at worst, spreading the costs so widely that their own constituents, while deriving the benefits, are saddled with only a fraction of the costs.

The electoral connection, then, implies two important facts of political life, one temporal and the other spatial. First, politically inspired real economic growth is targeted for election years. There is mounting evidence in this country and in others of a so-called political business cycle in which incumbent elected officials manipulate economic variables in phase with the political calendar. (The argument and evidence is found in Tufte 1978.) Whether they are successful or not, it appears that incumbents *try* to improve economic conditions just prior to elections. And recent evidence on American congressional elections shows that changes in real disposable income per capita in election years are strongly and positively related to changes in votes for candidates of the incumbent president's party. This, then, is a temporal fact of life.

Second, as Speaker of the House Thomas P. (Tip) O'Neill recently put it, "All politics is local." Political constituencies are geographically determined. Thus, while elected incumbents share an interest in timing economic benefits in even-numbered years, they are in competition with one another in terms of concentrating benefits on *their* constituencies, and either deferring or spreading costs.

These facts of political life deriving from electoral incentives have, I believe, some consequences relevant to the conflicts over energy policy. First, an energy policy based on no, or slow, economic growth runs directly contrary to the electoral interest of most incumbents. The preeminent "small is beautiful" politician, Governor Jerry Brown of California, has nevertheless devoted enormous personal energy and attention to economic development in his state. There is, then, strong electoral pressure for growth and an equally strong antipathy to no-growth.

Second, future generations have no direct voice in the electoral arena. Politicians responsive to electoral pressures are as present-oriented as their constituents are. This does not necessarily mean that the future is discounted entirely—after all, some voters are parents and grandparents, and most are not narrowly self-interested—but it does make it difficult for politicians to base their policy positions on a philosophy of intergenerational distributive justice heavily weighted in favor of future generations. President Carter's characterization of energy conservation and management as a challenge "the moral equivalent of war" has not exactly produced a tidal wave of self-sacrifice. I suspect former NFL football coach George Allen's slogan, "The future is now," would win more votes (if not more football games).

Third, economists are fond of distinguishing normal goods and services—called private goods, which are efficiently produced and allocated by competitive markets—from public goods. The latter have particular production or consumption characteristics that undercut the role that prices play in competitive markets. For some economists the defects of the price system are a sufficient justification for a positive governmental role in supplying public goods. (The classic statement is Baumol 1965.) While this is a controversial point among economists, the distinction between public and private goods is not. Nearly every undergraduate economics text makes at least a passing reference to national defense, interstate highways, and lighthouses, as classic public goods. What these texts often fail to note is that almost every example of a public good is, in fact, a *private good in production*. Thus a governmentally produced public good usually means contracts for private firms, profits for stockholders, employment for workers, and corporate, income, and sales tax revenues for states, counties, and municipalities. Similarly, a government-imposed regulation has profit, employment, and revenue effects felt at local levels. Elected officials are in a position to bribe their constituents, bringing home the "pork" and seeking relief for them from particularly onerous regulation in exchange for campaign contributions and votes. This form of electoral bribery is insensitive to, and therefore unaffected by, matters of optimal production and efficient consumption of energy. Moreover, it tends to reinforce the bias in favor of economic growth. I might also note that this phenomenon of electoral bribery is as true of Republicans as it is of Democrats, of conservatives as it is of liberals. In his 1972 campaign, for example, President Nixon stimulated the economy to the tune of a fourth-quarter growth rate better than 11 percent, and quickly turned the spigots off just after the election and began impounding funds.

Now, as Dr. Daly has noted (Chapter 2), socioeconomic and political trends and regularities are not necessarily ". . . constants of nature like the speed of light." The future need not be a projection of the past. I should like to underscore, however, that the electoral system has built-in incentives so that as long as incumbents seek to retain office, there will be, I believe, pressures for growth, for the distribution of federal largesse, and for relief from debilitating regulations. The "greatest good for the greatest number" is not, as Dr. Daly suggests, merely an invention of some defunct economist. Voters and consumers appear to subscribe to the principle, too. On balance, I believe, this severely damages an explicit, planned, no-growth scenario.

THE INSTITUTIONAL CONNECTION

If we turn from the electoral arena to institutional settings, the prospects for the no-growth point of view grow even dimmer. The reason is not so much that alternative scenarios are advantaged by the structural arrangements of the Congress and the bureaucracy. It is, rather, that these arrangements are not conducive to the long-term planning, regulation, and outright coercion required by the no-growth scenario.

At the most basic level our institutional arrangements are characterized by a separation of powers, in contrast to the fusion of powers found in many parliamentary systems. In effect, there is a constitutionally intended and built-in element of malcoordination. This, of course, does not mean that coordination is impossible, just unlikely. A national point of view is articulated by the president; various local points of view receive a sympathetic hearing in Congress; and the interests of functional groups—labor, producers, consumers, minorities—are represented and defended both in the Congress and in the bureaucracy. In order to effect change, coordination among the various branches of government and the interests that dominate them must be managed. This is complicated both by the heterogeneity of interests and by the absence of an obvious candidate for manager. The president is simply a representative of one special interest—a national or federal interest. Political parties in the United States, unlike some of their European counterparts, are principally electoral instruments, not policy organizations. And no ideology commands sufficient support to serve as a basis for a decisive coalition.

This malcoordination holds within each institution, too. For example, in the executive branch, while the president and his Department of Energy seek ways to reduce the demand for petroleum, the Civil Aeronautics Board is awarding additional routes to encourage airline competition, the Environmental Protection Agency is requiring cleaner-burning and therefore less fuel-efficient automobile engines, and the National Highway Traffic Safety Administration of the Department of Transportation is requiring safer and therefore often heavier and less fuel-efficient automobile bodies.

The structure of the Congress illustrates, in purest form, how institutional arrangements build in malcoordination. The House and Senate each have approximately twenty standing committees, 150 subcommittees, and a handful of joint and ad hoc committees. Committees, and increasingly, subcommittees, are masters of their specialized jurisdictional bailiwicks. Jurisdictional bailiwicks were created by the Legislative Reorganization Act of 1946, and have been modified only slightly in the intervening three decades. In the Senate, unlike the House, many energy issues are dealt with in a single, recently created, Committee on Energy and National Resources. Even there, however, there are separate subcommittees on energy research and development, public lands and resources, energy conservation and regulation, and energy production and supply. The subcommittees attract different kinds of members, often representing competing interests. The centrifugal forces are extraordinary. And if that were not enough, it should also be noted that rural electrification falls to the Committee on Agriculture, and military research and development, military stockpiles, and part of arms control are in the province of the Committee on Armed Services. Federal housing policies related to energy consumption are considered by the Committee on Banking, Housing and Urban Affairs; science, technology and transport are handled by the Committee on Commerce, Science and Transportation; resource protection and nuclear regulation are assigned to the Committee on Environment and Public Works; energy-related tax policy is handled by the Finance Committee. Other aspects of arms control and foreign economic policy are in the Foreign Relations Committee's jurisdiction; the Committee on Government Affairs deals with energy and nuclear proliferation; and Judiciary has antitrust and monopoly policy. Overlaid on that are congressional budget and appropriations processes that affect all authorizing committees. The House is even worse with more than two dozen standing committees and subcommittees with pieces of the action in the energy area.

In the light of the fractionalized form of our constitutional and institutional arrangements, how does anything get accomplished? The answer is complex. Occasionally, ad hoc coalitions are built within and between institutions for broad national policies. They tend to be impermanent and ultimately unstable as narrower, implacably competitive differences surface. More frequently, there are narrow special-interest coalitions among members of a congressional subcommittee, career employees of a governmental bureau, and interest group leaders. Called "cozy, little triangles," "policy whirlpools," or "unholy trinities," these coalitions are more durable because the set of interests is narrower and more manageable. Interest-group leaders float in and out of government, designing and managing government programs for bureaus that have obtained authorization and appropriations from their coalition partners in Congress. Thus, bureaus grow larger in authority and budgets; interest groups get the policies they want, and congressmen obtain benefits for their districts from the bureau and campaign contributions from the interest groups. Each coalition partner benefits. These coalitions of convenience are extremely durable, so long as they keep their demands focused. The result is a stream of partial policies, piecemeal in character, and often at cross purposes.

In sum, institutional fragmentation—both within and among institutions—combines with electoral incentives to diminish the government's capacity for long-term planning, systematic control over the economy, and influence over the private choices and behaviors of its citizens. Moral exhortation and symbolic acts may have some short-term effects, but they are reeds too weak to sustain the long-term objectives of a comprehensive plan like the no-growth scenario.

THE PUBLIC INTEREST CONNECTION

Before we answer the question posed by the symposium, "National Energy Issues—How Do We Decide?" there are two more fundamental questions to ponder: (1) Should we decide? and (2) Who are "we"? These questions, of course, do not lend themselves to scientific answers, but rather offer us the opportunity to articulate "orderly opinion," otherwise known as political philosophy.

Let me be a bit more specific. If a national energy program requires the coordination and coercion of a command-and-control administration of the economy, as I believe it does, then I am not at all sure that we *ought* to decide in any self-conscious sense, and certainly not in any comprehensive sense. Part of this answer derives from my own personal distaste for the governmental coordination and coercion which, I believe, will be required for comprehensive action. My personal opinions aside, I believe that a comprehensive national energy policy will have distributive characteristics that cannot be justified. It is no accident that the NAACP has, to the chagrin of its liberal allies, committed itself to the pro-growth strategy, just as it is no surprise that the Clamshell Alliance is hooked on the no-growth scenario. Each commitment reflects *private interests and values masquerading as the public interest.*

I urge you to beware of any argument beginning "The public interest demands. . . . " For some reason that eludes me, appeals to the polity to use the authoritative powers of the state always come wrapped in the clothes of the public interest. But in fact there is no such thing as the public interest; indeed, in my earlier comments I attempted to underscore the welter of private interests concerned with energy policy. In point of fact, theories of welfare economics and social choice do not provide us with firm criteria for resolving intragenerational or intergenerational distributive questions. In light of this, Dr. Daly quotes C.S. Lewis to the effect that "a dogmatic belief in objective value is necessary. . . ." The point is, however, that dogmatic or not, no objective value possesses either a logical basis or a compelling empirical foundation. Faith becomes the coin of such appeals and hence the attention which Chicken Littles and Dr. Panglosses receive.

In response, then, to the question of *whether* we should decide collectively, I believe the answer is still open. Of course, "nondecisions" have some of the effects of decisions: they have consequences for current and future distributional issues. Unlike positive decisions, however, they lack permanence, and the consequences that flow from them lack political legitimacy (though they need not be politically illegitimate, either). In this sense, nondecisions are noncommitments. In light of the large number of competing and equally legitimate private ends, both within and between generations, and the absence of criteria for weighing interests and trading them off, a case can be made for not deciding in any comprehensive fashion, for not having a national energy policy in this comprehensive sense.

One of the difficulties this approach encounters, and hence the necessity of asking the second question—who are "we"?—involves potential irreversibilities. Our electoral and institutional arrangements, as I have endeavored to describe, have a bias toward the present. I know of no way to give weight to the preferences of future generations, except as they are articulated by members of the current generation. Politicians today are responsive to questions of conservation, nondegradation, and restraints on development and growth, not because future generations command current votes, not because politicians regard their roles as trustees for the future as an irresistible imperative, but rather because *current* voters and institutional actors to whom politicians are responsive place weight on these matters. Whether this gives appropriate or optimal weight to the future is impossible to determine, if the question has any meaning at all. But it does give some weight, and this is accomplished by concerted lobbying and political pressure, just like the pleadings of any other special interest.

CONCLUSION

Let us sum up by repeating that our electoral process and political institutions are prejudiced against long-term, comprehensive planning of any sort. This is true whether the objective is rapid or sustained economic growth, on the one hand, or no-growth on the other. An occasional long-term strategy may be pursued by government, as in processing a war, but this generally requires yielding enormous powers to a commander-in-chief. More typically, our public policies are fairly short-term, tend to be synchronized with the political calendar, and are piecemeal in character.

The primary advantage of the malcoordination endemic to our policymaking process is its pluralistic nature, insuring that most interests have a voice, a hearing, and often a hand on one of the decisionmaking levers. The primary disadvantage, apart from the malcoordination itself, is the *pressure for decision*, particularly in order to have an answer to the electoral question, "What have you done for me lately?" The short-term, antiplanning bias probably means that the no-growth scenario is not politically viable. This does not imply that *all* of the values inherent in the no-growth scenario are impractical. Tactically, success requires that no-growth proponents shed their tendency to believe they have a monopoly on the public interest. In American politics, those groups that press for comprehensive programmatic objectives, arguing they constitute The Public Interest, all share one thing in common: they are losers! If, however, no-growth proponents roll up their sleeves, jump into the political trenches, and adopt the political tactics of other special interests, their point of view will carry some weight.

One concluding point: it appears to me that one valuable activity in which intellectuals and specialists in energy policy can engage is to press in opposition to what the late political scientist Schattschneider called "the mobilization of bias." The two biases I have identified— the pressure for decision and the pressure to trade off future welfare for present welfare—may have unhappy, irreversible consequences. Ironically, then, the best answer I can supply to the question "How do we decide national energy issues?" is to argue for delay in deciding publicly, in any comprehensive fashion at all. In effect, then, this suggests a governmental role in encouraging alternative energy strategies (perhaps even partially underwriting them in some instances), leaving research and development, when possible, and commercialization for sure, to the private sector. An early recognition of the incapacity of the public sector in the United States for long-term, comprehensive planning, and the superior capacity for "nicely calculated less or more" in the private sector, should dampen any enthusiasm for once-and-for-all formulations of national energy policy. It may be of some solace to know that in giving up the attempt at such comprehensive planning, we spare future generations our irreversible errors at a price: *they* will be burdened with having to decide.

REFERENCES

Baumol, William. 1965. *Welfare Economics and the Theory of the State.* 2nd ed. Cambridge, Mass.: Harvard University Press.

Tufte, Edward. 1978. *Political Control of the Economy.* Princeton, N.J.: Princeton University Press.

Discussion

Stephen Beckerman: As the only anthropologist here, I feel obliged to comment on two points. First, the question of cultural ends, which has been raised by at least three of the speakers: there have been those who claimed we needed them and those who claimed we could not possibly get them. I think it was a well-known biologist, Lawrence Slobodkin, who pointed out that every living creature plays an existential game with its environment, and the object of the game is simply to stay in the game. There is a basic selection principle operating on all societies as well, and all societies which survive (which are all the societies we can observe) have survival as one of their goals either explicitly in the short run, or implicitly in the long run, as a sort of cultural metalogic.

The second comment has to do with Dr. Lave's statement about the rise of the standard of living with the general rise of the consumption of energy. It may or may not be true that, in any particular case, a certain group of neolithic villagers had a higher general standard of living than a certain group of hunters and gatherers. I can produce a number of counterexamples. The Dobuans are thoroughly miserable, although they are agricultural, and Congo pygmy hunters are generally fairly happy and well off. When we move to stratified, complex societies, it is probably the case that the standard of living of people who have the *highest* standard of living in a particular society goes up as the energy consumption increases. It is not true that the average standard of living goes up. When England was recruiting soldiers for the Boer War, she must have had the highest per capita energy consumption in the world, and certainly the English aristocracy and the English merchant class had the highest standard of living that the world had ever seen. But well over one-quarter of the men who were drafted had to be rejected on the grounds that they were so weak, so malnourished and physically debilitated from the conditions under which they lived, that they were not good for cannon fodder. It is not necessarily true that the average standard of living goes up as energy consumption increases.

Gene Rochlin: I would like to start with a piece of data, something I think has been sadly lacking in the preceding papers. Last year, the United Kingdom issued a report called "The Future of the United Kingdom Power Industry" in which it was pointed out that the United Kingdom had so over-planned capacity that no—zero—new power plant orders were expected for perhaps a decade or more, starting in the early 1980s. That nation's worry over its over-planned capacity was that by the 1990s it would no longer have an electricity generating manufacturing industry.

The report identified four alternatives: (1) let it collapse and simply not care—let these things work themselves out; (2) force industrial mergers with other European countries in order to maintain the business; (3) build power plants the nation does not need to keep the industry going; and (4) stimulate the economy in Britain and increase energy consumption so that the plants are built and would be needed.

The reason I am bringing this up is to illustrate the nonseparability of variables. The problems we talk about here cut across many things. Each one of the speakers has, to a different extent, taken one cut via his own specialty and acted as if that tail was the whole dog— perhaps the dog is made up of nothing but wagging tails!

The specific point I wish to raise concerns the role of planners— the role of "we," which is where we started this morning. If you look at Dr. Malès' U-curve, I would point out that if you underbuild, there is a push on supply, whereas if you overbuild there is a pull on demand. Notice that the way that curve works is, in *either* case, to increase either consumption or supply. There is no way, in these three variable models, to obtain a decrease. It would appear, then, that planning becomes the only truly independent variable in the model; then the planners themselves become "we." If there is intensive lobbying about planning and models, it is because one fears that the assumption that other variables are independent of planning is wrong.

Treating people as if they were individual units and the social costs and benefits as if they accrued to individuals is too limited. We have heard very little so far about the collective benefits and the collective nature of society except in the last paper. Energy may be only 7 percent of the GNP, but its effect is collective in many ways.

Also, the people who say, "Let's not make energy seem so important; get on with the energy planning itself, and forget all this stuff about life-style and so on. . . ." are throwing away the major piece of sociological data available! From a sociological point of view, that *is* a data point; it is important because it is believed to be important. Perceptions are data in many social science fields.

Marc Roberts: I have two points to make. First, Dr. Malès' comment that uncertainty leads to higher levels of optimal capacity expansion may be true, but the really critical issue is exactly how you define the mechanics of the model. It could lead to an expansion in the optimal reserve capacity by 1 percent, in which case it is true but uninteresting.

My criticism of Dr. Malès' analysis is that it did not present enough data to judge and evaluate whether this was a trivial or significant effect. I was stunned to hear him say that 15 to 20 percent reserve capacity margins were much too low, in light of the fact that Pacific Gas and Electric operates on 12 and 13 percent and many large systems operate well below the 15 to 20 percent level. I think Dr. Malès needs to give us more data in order for us to judge whether it is really an important problem or not.

My second point concerns Dr. Daly's notion of "sufficiency." Presumably, he intends its definition to come from that "clear and certain knowledge of objective values" he spoke of. I agree with the last speaker that no statements about the world can imply statements about values. There is a distinction between facts and values. I see no way, therefore, for there to be such a thing as "objective" conflict resolution of conflicts in values. So when Daly asks for "objective values" as a way to build social consensus, he is invoking a process I find literally incomprehensible. He talked of a need for faith in "objective values," and perhaps he means we need to found "objective values" on faith. That is at least a process that I can understand. But it does not seem likely that a common "secular faith" is a practical basis on which to build current social consensus.

Because I believe that values are ultimately a matter of "faith" and not "objective," part of the difficulty that I see in Dr. Daly's whole presentation is that it leaves out the dynamic and political process aspects of any kind of value-building process. It ignores the role of leadership and vision in any sort of value-defining process in a society. Simply because processes are not objective in the scientific sense does not mean that they do not occur.

It does seem to me that it is very unlikely that we are going to agree to some "objective value" basis of social policy as long as we maintain any kind of democratic basis of political action. Furthermore, I agree with the comments of other speakers that, as best we can perceive from the current consensus of values in the society, it is one that would not make Dr. Daly very happy.

Jules Josephson: My remarks are directed to Dr. Shepsle. You described our political process and drew some conclusions from it. Specifically that "there is a strong electoral pressure for growth and an equally strong antipathy to nongrowth." In addition, you find that "if we turn from the electoral arena to institutional settings, the prospects for the nongrowth point of view grow even dimmer." You sum this up by pointing out that "our electoral process and political institutions are prejudiced against long-term, comprehensive planning of any sort." Presumably then, our economy would not be too receptive to planning to take the soft path of energy development.

The Soviet Union has opted to satisfy their energy needs by planning a large investment in fossil and nuclear power with little emphasis on solar, geothermal, or wind power. Because the Soviet Union has an almost totally planned economy, and in addition, since their leaders need not be responsive to current voters and other institutional actors, do you think that they could plan and carry out an alternative energy path unhampered?

Kenneth Shepsle: One of the distinctions that I left out of my planned comments was to compare our system of government, one that is broadly characterized as a "separation of powers," with the notion in the British and other parliamentary systems of the "fusion of powers." I was afraid that "fusion of powers" would be misunderstood here! Anyway, the capacity to conduct planning of any sort is greatly debilitated by a separation of powers, which is a concomitant of our constitutional arrangement. I do not think, as a consequence, that a separation of powers is a bad thing; rather, what it does is relegate many planning functions, by default, to the private sector.

In the public sector, under a political system that does not have a command and control effect on its economy, most of the planning will not get done publicly. Most of the results in the public sector will be piecemeal, and often at cross purposes.

Herman Daly: With respect to Dr. Lave's comments, I'll take a number of them to heart. I want to thank him for offering me a place in a democratic society as one who performs the service of upsetting inertia by means of shrill discourse. Other commentators have not been so kind; they have read me out of the democratic society entirely, because I did not advocate the majority view, which I find is rather a strange attitude for democrats.

I did not really talk about the idea of a steady-state economy. But I did mention it peripherally, and so the question was raised by Lester Lave, "What is a steady-state economy?" Certainly, that is a proper question. The notion of a steady-state economy is defined in physical terms: constant population and stock of physical artifacts maintained by a minimal rate of entropic throughput, or metabolic flow, from the environment and back to the environment. There is still room in such a system for qualitative change, for technical improvement; in fact, the steady state becomes the norm. Growth, then, is seen more as a temporary process of moving from one steady state to another steady state. If you invent new technologies which make it both desirable and possible to move to a higher level of population and energy consumption, and so on, then you can do that. What you want to avoid is to be pushed to a new level of growth, to a neolithic society when your technology is still back in the hunter-gatherer stage. Let the technology improvement occur first, then you can grow into that expanded niche. Don't grow into the niche and then have faith that somehow the technology will come along. That is a distinction.

I want to correct the impression that I reject economic analysis; as an economist, I would prefer not to reject economic analysis. In other words, there are costs to growth; there are benefits to growth. There are marginal costs to growth; there are marginal benefits to growth. Growth should cease where marginal costs equal marginal benefits. Let us talk about an optimum level to attain. Let us not think of growth as something that can somehow escape the economizing calculus.

As for Dr. Shepsle's comment, that at present there is electoral pressure for growth, and there is no great consensus behind nongrowth—that is certainly true. I am not planning to run for president; I do not think that at the current time there is political support. That is just to say, we are at the beginning of a process; it has not gone very far. As for the notion that coercion is implicit in nongrowth, and that when you talk about a nongrowing economy, you are assuming Draconian measures that take away an individual's freedom, I think that is not the case. To make a proper comparison when talking about aspects of freedom, we should *not* compare freedom as it has existed in the economy of the past, with freedom as it is likely to exist in a growth economy of the future. I do not think that it is going to be easy to maintain the degree of freedom that we have had in an era of high growth. I think that an economy which attempts to force growth in the face of rising natural and social resistances is probably going to require more coercion than an economy that accepts limits and tries to live within them.

As for Dr. Roberts' comment from the floor regarding objective value, what I meant by that is not that values are to be found in the physical world by objective scientific procedures. I meant to distinguish the notion of true objective value from subjective values in the sense of tastes. I do not see any point in considering values subjective in the sense that "You have your preferences and I have my preferences . . . why should we discuss it?" That is taste. Values are things that are much more important. I have my values and you have your values; that is a serious matter, but if we differ, that does not mean that we have to start shooting at each other. But we have to take it seriously to work out some sort of consensus. In that sense, it is what I mean by appeal to objective value. To put it another way, a dogmatic belief in objective value is to policy and decisionmaking as a dogmatic belief in an orderly universe is to research in physics. If one rejects the notion of objective value and the concept of the public interest, as does Dr. Shepsle, then it seems rather a waste of time to attend a conference on "National Energy Issues—How Do We Decide?" By that hypothesis, there is no real decision, only a fight among private interests. Argument would become a tool of interest, not a process of collective reasoning together in search of objective value.

René Malès: First, I understand that Dr. Lave, in a whisper at least, agrees with me that Say's Laws may not be correct. He may want to defend himself!

Then, to the comment that Dr. Rochlin made from the floor with regard to the pull on demand created by excess capacity, I question whether that is a fact. The example he gave of the power industry in the United Kingdom is interesting, but reflects other social considerations and not the pull on demand. In fact, my whole experience in developing new technologies is that it is very difficult to get consumers to buy. My friends in the manufacturing field keep telling me they spend a great deal of their resources in marketing, just to cause people to desire the "better mouse trap." It is not enough just to have spare capacity to create a demand pull.

Second, on the point that Dr. Lave made that "a BTU is not a BTU is not a BTU" à la Gertrude Stein, he is perfectly correct. However, correcting for those statistical deficiencies to the extent possible reveals that there still remains a difference in the energy intensity during this period of time, although not as great as is shown in the charts. The concept of the different values for different kinds of BTU's is terribly important. Not making such differentiations has led us into all sorts of stupid policy decisions. Finally, I accept the point he was making on the difference between need and demand.

With regard to Dr. Shepsle's comments, let me highlight: delay is a very costly factor, so costly that it may not be acceptable; but the cost may not be visible. Maybe that is the reason it is accepted. As you probably know, the cost of a delay for starting up a nuclear plant like Diablo Canyon or Seabrook runs on the order of $1 million per day to $1 million per week, depending on who is measuring, and how one measures cost. In any case, it is a very substantial sum. Unfortunately, consumers have those figures obscured from them. It may be a decision that they would not like to make, but they do not even know they are making it.

Finally, on Dr. Roberts' comment made from the floor on the question of reserve margin, he is correct that Pacific Gas and Electric operates on a very low reserve margin. One reason is that they have a very strongly interconnected system. The second reason is that their base capacity, about 50 percent, comes from hydrocapacity. This gives them quick start-up capability. That is not true of the typical system. Commonwealth Edison also used a 12 percent reserve margin for years. Recently, they increased it to 14 percent because of higher outage rates of their larger fossil and nuclear units. But the typical system in the United States uses a standard margin of a 15 to 20 percent. Incidentally, this is an area in which significant improvement can be made from a technical point of view.

The other point that Dr. Roberts made, whether uncertainty creates only a small amount of additional reserve margin, is also important. It amounts to 5 to 10 percent of additional reserve margin requirement in the systems tested with the degree of uncertainty used.

Alvin Weinberg: Well, time has run out for this session, illustrating what I think, in a way, is one of the most curious and perhaps deepest aspects of the whole energy issue. That is, that of all man's resources, we think of energy as the one we have to husband the most, but—think about it a bit—*time* is equally important. There is a growing literature about the trade-off between time and energy, illustrated during the discussion and questions evoked by the remarks of Dr. Malès with regard to outages.

Intermittence of energy supply, which means the capacity to allocate one's time as one wishes, is very fundamental to the whole argument that was raised by Dr. Daly: the soft path versus the hard path. I would expect that future discussions will focus on how these two different paths affect what many economists and sociologists consider to be man's most precious resource, time. *This* is the ultimate issue, but clearly, we are not going to settle it in the half-minute we have left.

Session II

Decision Criteria for Health, Safety, and Environment

Among the central questions concerning the use of nuclear power in general and plutonium in particular is: How safe is it? The question is essentially a quantitative one which could be answered if we had adequate knowledge of the kinds of accidents that might occur and their chances of occurrence, as well as adequate knowledge of the health and environmental effects of radioactive materials. Although efforts to increase understanding of these questions have been underway since the earliest days of nuclear power, much still remains to be learned. The current state of knowledge and how it can be used to establish criteria for plutonium use is the subject of much of this session. By its nature, the subject is technical and quantitative, but this does not mean that it is free of uncertainty. In this discussion, for example, uncertainty is illustrated by the disagreements about the interpretation of the data, which carry over into the post-symposium discussion in the appendix. The session winds up with consideration of the social aspects of the health and safety issues and public attitudes toward the risks involved. In the long run it is these sociopolitical factors that will determine the acceptable criteria for health and safety.

Session II

Opening Remarks by

Chairman: *David L. Sills*

The topic for this afternoon is "Decision Criteria for Health, Safety, and Environment." If you look at that phrase hard, as I have been doing, it turns out to be virtually the agenda of the entire world for the next several decades. Clearly, we have in the use of plutonium a technology that requires both evaluation and the development of decision criteria, and this has to be done with skill, subtlety, and political sensitivity. You may remember that ten or fifteen years ago, there was a great decision criterion called "Better Red than dead!" We have become far more sophisticated in our political analysis since then.

The papers this afternoon are, from the point of view of a sociologist, technical; that is, they are papers on the technical problems of evaluating toxicity and the safety of nuclear power in general. However, I do not think that this makes them any less important or less global. We are going to have three discussants who will, I believe, point out that plutonium has both the kind of toxicity that we are going to hear about and a certain element of political toxicity!

✳ *Chapter 5*

Setting Safety Criteria

Norman C. Rasmussen

Since the beginning of civilization each new technological development has brought not only benefits but certain risks. In adopting new technologies societies have either implicitly, or sometimes explicitly, decided that their benefits outweighed their risks. A review of history indicates that mankind's drive to improve living conditions has led to the belief that technology provided mostly benefits and that technological development was desirable at as fast a rate as possible.[a] This attitude is still quite prevalent around the world. In recent years, however, some of the more affluent societies, such as ours in the United States, have experienced a growing questioning, by some people, of the desirability of continued technological growth. Each new technology is asked to demonstrate that its benefits outweigh its risks and to prove that it presents an acceptably small risk of damaging either the public health or the environment. Because of the increasing scale of industrial activities, I am sure most of us feel that it is prudent that the risks of each new activity be carefully examined. The process of carrying out this examination has developed in such a way, however, that it threatens to paralyze a number of industries that provide goods and services essential to advanced societies as they now exist.

In the United States, and many other countries as well, the response to legitimate concerns of the public about technological risk has been to create regulatory bodies to assure the safety of the public

---

[a] It should be noted that for centuries there has been questioning of the benefits. See, for example, *De Re Mettalica*, Agricola (1556) for questions about the developing of mining.

and the environment. Regulatory bodies have been created by the government gradually as technologies seemed to present significant risk. In the United States, we find regulatory bodies responsible for mine safety, consumer product safety, nuclear safety, food and drug safety, occupational safety, environmental protection, and so forth. Not surprisingly many confrontations have developed between the regulator and the regulated. Properly controlled, this would probably be a healthy situation. Unfortunately, these confrontations are threatening to paralyze further progress and development in some industries, and to such an extent that national goals are being thwarted. For example, there can be no doubt that the United States is facing serious economic problems as a result of importing annually about $45 billion worth of oil. Yet the opening of new coal mines, the drilling of off-shore oil wells, or the building of new nuclear power plants, are being subjected to ever-increasing delays by regulatory actions and legal processes. The drug industry offers another example. During the 1970s, the annual rate of introduction of new drugs has dropped dramatically in this country, despite the fact that the United States leads the world in investment in medical research. There are numerous examples of effective medicines widely used in other countries that are still not approved for use in the United States. Many have concluded that this can be directly attributed to over-restrictive regulation of the industry. Why do we have these problems?

Clearly, there are a number of contributing factors in each case, but in my opinion, the most important underlying cause is that the regulators have either unreasonable criteria or poorly defined criteria for what is an acceptable level of risk. The regulators tend to reflect society's desires, of course, so this is not an indictment of them, but rather an indication of the dichotomy that exists within society between reducing risk and increasing cost. The regulator is faced with the unenviable task of striking the proper balance. What the regulator would like is a clear statement by society of what level of risk is acceptable. This would please the regulated as well, since one of their main concerns is the uncertainty caused by what appears to them to be the regulators' changing definition of acceptable risk. Despite the desirability of having well-defined criteria, setting them remains a difficult task.

I have been asked to discuss safety criteria here today. Since my involvement with the Reactor Safety Study (U.S. Nuclear Regulatory Commission 1975), I have thought about this subject often. The conclusions I have reached seem to be generally similar to those held by others who have written on this subject.

I will start this discussion of safety criteria by mentioning a few cases where criteria have been defined and have worked well. I will then present some examples of criteria that seem to be failing badly. Finally, I will discuss the case of plutonium and the possibility of developing acceptable safety criteria for its use.

EXAMPLES OF SAFETY CRITERIA

Let us look at a few examples of safety criteria that have been set and appear to have achieved their desired goals.

Automatic Landing Systems for Aircraft

Some years ago, the United Kingdom developed a fully automatic landing system for commercial aircraft (Warren 1973). Once the principles of the system were demonstrated, the civil authorities responsible for aviation safety were faced with the question of defining what failure rate of this system would be acceptable for commercial aviation use. The authorities decided that the rate of fatal landings (landings resulting in one or more fatalities) must be demonstrated to be less than 10^{-7} per landing (one in ten million). They reached this criterion by noting that the then current rate of fatal landings was 10^{-6} (one in one million) per landing in commercial aviation. It was expected that the introduction of the automatic system, qualified to 10^{-7} per landing, should result in an increase in overall passenger safety.

It should be noted that this criterion alone would have been of little use if the regulatory authorities had required twenty or thirty million landings to demonstrate statistically that the criterion had, in fact, been achieved. Instead it was agreed that reliability analysis could be used to verify that the level has been achieved.

Using the defined reliability goal, engineers designed a system to achieve it. The system was bench-tested to demonstrate that the design goal had been achieved. The tests resulted in numerous design changes. Once the bench tests indicated that the desired reliability had been achieved, a number of systems were installed in instrument aircraft for flight tests. Using the operating characteristics from the bench tests, the reliability analysts predicted the likelihood of deviations from the normal landing pattern. The results of many hundreds of actual flight tests were then used to verify the reliability calculations. Once the observations and calculations were in agreement, the system was accepted as commercially qualified.

In my opinion, the foregoing criterion is the simplest type of safety criterion to set for several reasons. First, the new system is being substituted for an already accepted system. Second, the new system meeting the designed criterion offers an increase in passenger safety. It also offers an economic benefit to the airline operators by enabling their planes to land in bad weather. Another factor that should be noted is that failure of the new system has the same consequences as failure of the old system. Thus, in considering the comparative risk, one need only be concerned with the probability of failure for the two cases. Unfortunately, there are few examples where things are so simple.

High-Integrity Protection System

The second example I have chosen is the design of a safety system for a new chemical plant (Stewart and Hensley 1971). This system was designed to shut off the feed materials to a chemical reaction if the mixture should approach the explosive limit. The chemical company (Imperial Chemical Industries) had to develop a safety criterion that would be acceptable to the employees' union. The company proposed a hazard rate of 3×10^{-5} per year. The hazard rate, or rate of system failures that potentially threaten the safety of the worker, was chosen because it was 0.1 the average hazard rate of all company operations. This basis for design of the new safety system was accepted by the employees' union. The criterion was practical only because it was agreed that the reliability of the system could be demonstrated by reliability methods, as in the previous case. Again, the criterion was acceptable because it represented an improvement in safety over an already existing practice. The 3×10^{-5} per year criterion led to a very expensive, triply redundant set of shut-off valves. Despite its high capital cost, the company has found the system to be paying for itself because its high reliability has reduced the number of spurious plant shutdowns experienced with the previous safety system.

The foregoing criteria apply in cases where the people exposed to the risks were receiving direct benefits from the activity. Moreover, we may presume that the exposure to risk was voluntary in the sense that people might have chosen not to fly in a plane or work in a chemical plant. It should be recognized that people are sometimes killed by falling aircraft or by chemicals accidentally released from a nearby plant, and to them, the risk is involuntary. As is often the case, both voluntary and involuntary risks are involved in setting criteria.

Dikes in the Netherlands

As another example, let us consider the case of the dikes in the Netherlands. You will recall that several decades ago the dikes, which protect nearly one-third of that nation, were breached by an extremely severe North Sea storm. The Dutch government was then faced with deciding how high to rebuild them. It was decided that the dikes should be high enough to protect against a North Sea storm with a return frequency of one in 10,000 years. I do not know exactly how this return frequency was chosen, but I suspect it was chosen for the following reasons:

1. The cost of the project was economically feasible for the Dutch government.
2. The failure probability of 10^{-4} per year could be defended as quite rare in that such a storm would have been expected about once in all recorded history.
3. The new dikes would be higher than the previous ones, thus offering an improvement in safety.

To summarize, the preceding risk criteria had several features in common. (1) They appeared to reduce risk over current practice. (2) They accepted reasonable methods of proof that the criteria had been met. (3) The fixes required by the criteria did not lead to severe economic penalties to the activity. (4) They were defined on a logical, understandable basis.

EXAMPLES OF UNSUCCESSFUL CRITERIA

The Delaney Amendment

In 1958 the Congress passed the now famous Delaney Amendment, which said in part: "No (food) additive shall be deemed to be safe if it is found . . . after tests which are appropriate for the evaluation of the safety of food additives to induce cancer in man or animal . . . ." When passed this ruling seemed reasonable; after all, who wants a carcinogenic substance added to his or her food? The problem was that the law as interpreted by the Food and Drug Administration said that no risk of cancer, no matter how small, was acceptable. In order to determine whether a substance was carcinogenic, tests were developed that consisted of feeding massive doses of the substance in question to colonies of cancer-susceptible mice.

The result of the Delaney Amendment is that cyclamates, Red Dye #2, and a series of other substances have been banned. As required by the law, in all cases banning was done with no consideration of whether the banned substance might have a benefit that outweighed the risk it presented, and with no consideration of the actual risk level. The matter came to a head with the banning of saccharin. Some studies in mice had indicated that massive doses of saccharin might cause bladder cancer, but it was clear that at the rate of consumption of normal users of this sweetener, the risk of cancer was exceedingly small. Despite the fact that many observers believe that the use of saccharin reduces the probability of heart disease, the number one cause of death in the United States, the Food and Drug Administration proposed the ban. A zero-risk criterion is an unreasonable criterion, and in the end, will create problems and not solve them. This is apparent in the current saccharin debate.

Nuclear Power Safety Criteria

One of the most controversial areas of regulation today is the nuclear power industry. In approving any project, the Nuclear Regulatory Commission (NRC) states that the project must meet its criterion of "causing no undue risk to the public." A criterion like this is no criterion at all, of course, because everyone is likely to have his or her opinion of what constitutes undue risk. The licensing hearings have clearly demonstrated this. Let us consider a few specific examples from the nuclear case.

For a number of years, the NRC has struggled to establish an acceptable level for the low-level emission of radioactivity from normal plant operations. It finally agreed upon the principle of allowing a level "as low as practicable" (ALAP), which means that the best currently available technology should be employed to reduce emissions. As technology has improved, the ALAP criterion led to more and more expensive control equipment. The problem of high cost led to the suggestion that the safety criterion be respecified on the principle of "$1000 per person rem." Thus, the plant operator would be required to install equipment as long as it cost less than $1000 to reduce the population exposure by 1 person rem. Although in principle such a criterion may seem like the right way to deal with this problem, it stirred up considerable controversy because it was not consistent with the levels of risk from other sources. The reason for the inconsistency is that it is generally agreed that a population dose of about 10,000 person rem will produce one additional cancer in a ten to forty year period following the exposure.

The criterion of $1000 per person rem implies that the value of a human life is:

$$1000 \ \frac{\text{dollars}}{\text{person rem}} \times 10{,}000 \ \frac{\text{person rem}}{\text{fatality}} = \$10{,}000{,}000.$$

This value is certainly not consistent with capital investment in other safety systems to save a life. For example, does anyone doubt that if we had very strict enforcement of traffic laws and the best safety design of our highways, that we could save 10,000 lives per year? We have achieved a large fraction of that by having a poorly enforced 55-mile speed limit. If, in fact, a life is worth $10 million to society, we should be willing to invest $100 billion more annually into highway safety to achieve a reduction of 10,000 fatalities. It seems clear that our society does not value the lives lost on the highway at anywhere near $10 million per life.

The present nuclear regulations call for a specific annual radiation dose at the site boundary that is equivalent to about 5 percent of the natural background radiation. People living any distance from the site boundary, of course, get a substantially smaller dose. Although this seems to be a very conservative criterion, it does have the advantage of not changing as the control technology changes. I believe that the great controversy in this area has been in large part due to a poorly defined criterion, followed by a proposed criterion that was unreasonable.

A second issue that has created considerable controversy in the nuclear industry is the hypothetical accident sequence known as anticipated transient without scram (ATWS). This refers to hypothetical accidents in which the reactor shutdown system fails to operate. In the early 1970s, the Atomic Energy Commission (now NRC) issued a technical report (U.S. Atomic Energy Commission 1973) in which is stated that the probability of a serious ATWS event should be 10^{-7} per year or less. The report suggested further than this degree of reliability could be achieved by a redundant and diverse shutdown system. Members of the industry suggested designs of systems they believed would achieve the desired reliability, only to learn that the NRC staff would not accept their analyses of reliability. It appeared that although the NRC would accept 10^{-7} for the failure rate, none of the analytical methods available for calculating system failure rate was acceptable to the NRC as proof. This issue continued to be debated until the NRC released a position paper (U.S. Nuclear Regulatory Commission 1975), proposing a criterion of 10^{-6} for a serious ATWS event. It is not clear yet what methods of proof will be accepted. Clearly, in this case most of the difficulty was created because there was no accepted way of proving that the criterion had been met.

CRITERIA FOR PLUTONIUM

Today we as a society are debating whether to develop a nuclear industry based on the use of plutonium. In addition to being a very toxic radionuclide, plutonium can be used to make nuclear explosives, and so it is perceived to be a substance whose use in commerce presents significant risks. Of course, it can provide substantial benefits: breeder reactors using depleted uranium as fuel offer the possibility of an energy-producing system with almost limitless fuel supply. For example, the approximately 170,000 tons of depleted uranium in government storage has the capacity to produce as much electricity as the entire known coal reserves of the United States. That uranium is already mined, above ground, and in drums. However, the use of this nuclear fuel cycle implies the commercial processing of substantial quantities of plutonium and it is extremely difficult to set acceptable safety criteria for these operations.

As shown, one way for safety criteria to gain acceptance is to reduce the risk of a currently accepted process. In this case, plutonium could substitute for the burning of hydrocarbons for electricity production and thus eliminate the risks associated with their mining, transportation, and combustion. It is extremely difficult to make meaningful comparisons between the risks associated with the use of hydrocarbons and those associated with the use of plutonium in generating electricity. How does one compare the risk of proliferation, the possible, but unlikely, meltdown of a plutonium-containing core, and the long-term risks of waste disposal, to the risks of climate modification by CO_2 emission, health effects of SO_2 and NO_x, and the impact of mining and transport of large amounts of coal? How does one estimate the increased risks of global conflict if failure to exploit the nuclear option leads to increased pressures on oil supplies? If the problem is viewed in this global manner, I suggest, none of us will be able to make a comparison that will be convincing enough to set a widely accepted set of safety criteria for plutonium use.

It may be possible to use the principle of defining acceptable risk criteria for various parts of the plutonium cycle by arguing that the chance that the plutonium cycle will produce a given consequence is small, compared to some existing way of producing the same consequence. As an example, consider the very difficult proliferation issue. Plutonium is not the only material for making an explosive; both Uranium-233 and Uranium-235 will also do nicely. Uranium-235 is widely available around the world as the 0.7 percent component in natural uranium. At present, probably the simplest way to get Uranium-235 in high enough concentration for making an explosive is to use centrifuge separation. A reasonable criterion for the plutonium cycle might be that the plutonium always be in a form that would make it more difficult to obtain in weapons-usable form than obtaining Uranium-235 from natural uranium. I do not mean to suggest that such a criterion would be easy to develop, but certainly it would be easier than the more global comparison noted above. Dr. Rowen will have even more to say on this subject in his presentation (Chapter 11).

Another question that is paramount in the nuclear controversy is that of long-term storage of radioactive waste—the foremost issue in the nuclear debate today. How can we estimate the risk posed by a waste deposit 10^5 or 10^6 years in the future? How can we ensure the integrity of a storage site for such an extended period? The answer is that we cannot really predict with confidence what conditions will exist so far in the future. Hence the setting of acceptable safety criteria seems impossible. I would suggest that we consider that the earth already contains many radioactive deposits containing long-lived alpha emitters of the type present in reactor waste. In fact, numerous studies have shown that the radioactive character of reactor waste after three to five centuries is similar to that of uranium ore body. Thus, the criterion could be based upon keeping the risk smaller than that presented by natural uranium deposits. For example, the waste should be placed in a region less accessible to ground water and in a form less soluble than typical uranium ore bodies. I believe that it is well within the capability of today's technology to demonstrate that such a criterion has been met.

In addition, it should be possible to pick a site where we have a high degree of confidence that the formations will be stable for the five centuries it takes for the shorter-level fission products to decay. Such criteria have a simple logical basis that could prove acceptable to a majority of people.

The question of power plant safety may be the most difficult. Since today's water reactors contain almost as much plutonium as a breeder reactor, the problem is little different from the one we already face. Although overall risk assessments of the type carried out in the Reactor Safety Study (U.S. Nuclear Regulatory Commission 1975) are a very valuable exercise in developing an understanding of the safety issues, I think it is fair to say that the uncertainty in the absolute value of the risk level is large enough that, until these uncertainties can be reduced, they cannot provide a widely accepted basis for verifying that a safety criterion has been met. It does appear that comparative risks determined by these methods are much more accepted in the scientific community.

The NRC has been able to take advantage of the foregoing methods to demonstrate that certain designs and procedures are safer than others. A number of safety issues have been downgraded in importance by showing that they are small compared to other issues. The use of reliability methods in a comparative way has helped the NRC resolve a number of otherwise thorny issues. Thus trivial issues are eliminated or played down and attention is focused on the dominant contributors to the risk. The problem remains, however, of deciding when the dominant contributors to the risk have been dealt with in an effective way.

Since the uncertainty in our estimate of the reactor accident risk lies mostly in estimates of the probability of the various accidents rather than the consequences, one can be confident that the risk is less than that calculated from an accident-free record of about 300 plant years. Using this approach, a Ford-MITRE Study (Nuclear Energy Policy Study Group 1977) concludes that today's nuclear plants offered no greater risks than coal. The study also carefully points out that this was an extreme upper-bound estimate of the nuclear risks.

Good criteria are certainly not easy to develop. To be useful, they must first and foremost be derived in a simple, logical way that can be understood by the interested parties; and they must include an assessment of both benefits and risks. They must set risk levels that are somewhat lower than the currently existing risks. It is essential that there be accepted ways to demonstrate that the activity meets the criteria. Of course, they will be nothing more than a stop order on the activity unless they can be achieved within existing economic constraints. As simple and obvious as statements such as these seem to be, the examples given indicate that they are often not comprehended by regulatory systems.

REFERENCES

Agricola, G. 1556. *De Re Mettalica.* Book 1.

Nuclear Energy Policy Study Group. 1977. *Nuclear Power Issues and Choices.* Cambridge, Mass.: Ballinger Publishing Company.

Stewart, R.M., and G. Hensley. 1971. "High Integrity Protective Systems on Hazardous Chemical Plants." Paper presented at the European Nuclear Energy Agency Committee on Reactor Safety Technology, SRS/COLL/303/2. May.

U.S. Atomic Energy Commission Regulatory Staff. 1973. Technical Report on Anticipated Transients without Scram for Water-Cooled Power Reactors. WASH-1270. September.

U.S. Nuclear Regulatory Commission. 1978. Anticipated Transients without Scram for Light Water Reactors. NUREG-0460. April.

_____ . 1975. Reactor Safety Study: An Assessment of Accident Risks in United States Commercial Nuclear Power Plants. WASH-1400 (NUREG 75/014). October.

Warren, D.V. 1973. "Safety Assessment of Systems for Landing Aeroplanes in Bad Visibility." Unpublished. October.

Chapter 6

Plutonium in the Environment—Individual and Population Risk

Gordon Burley

The topic of this symposium—National Energy Issues—is very broad. Our field of view has been narrowed, and I think usefully so, by emphasizing the importance of plutonium in such considerations. From the viewpoint of an environmentalist and because of my association with a federal regulatory agency, I can, perhaps, bring to this discussion a perspective different from that of my colleagues on this panel. My paper outlines the rationales for protection of individuals and populations and indicates the experience of the Environmental Protection Agency with development of radiation protection guidance for persons exposed to plutonium in the environment.

Criteria for minimization of risk and rationales for protection are obviously interrelated and serve the same objective. There are, however, several different types of rationales for protection and it is useful to classify them and consider them in more detail.

The first category of rationales for protection is that of engineering criteria, which vary in level of stringency. The most restrictive level requires "best available technology," the best engineering know-how; it affords few options. The next level, "as-low-as-practicable" (ALAP), introduces an element of judgment; it is followed by the very similar "as-low-as-reasonably-achievable" (ALARA) criterion, which introduces the function of economics and the balancing of costs and benefits.

A second category of rationales of protection is that based on risk, both absolute and relative. For radioactive materials, these rationales are based primarily on complex correlations of absorbed dose and adverse health effects. Our knowledge is sadly lacking in most areas concerned with health effects, but because radiation, and especially plutonium, has been better researched than most other carcinogens, we can make reasonably definitive judgments.

The last category of rationales of protection comprises judgments concerned with the larger perspective of societal impacts. This perspective includes the balancing of the costs and benefits of an activity, especially from the broader viewpoint of its impact on the population as a whole. It may also include a comparison of a product or activity with others in terms of the risk and benefits, either on a direct basis or on a relative scale. Costs are intended to include both the monetary and nonmonetary (such as environmental) impacts. Quantification of societal costs and benefits in such a context is obviously difficult and often controversial. Moral judgments that consider the effects of current activities, not only on the present, but also on future generations, are involved in such a decisionmaking process. For example, plutonium–239, which has a radioactive half-life of about 24,000 years, will not completely decay in more than one-quarter million years. Therefore, the consequences of decisions made today go beyond terms of human existence and are in the realm of geological ages. The ramifications are truly enormous.

We all know that there is an ongoing argument in this country with respect to nuclear power, especially concerning whether to build breeder reactors. An argument arises when there are different perceptions of the risks and benefits associated with an activity. I will confine my remarks primarily to the risks.

Plutonium was first produced for the weapons program. Its history has to some extent influenced the attitude of society toward this element. As a result of the atmospheric tests of the 1950s and early 1960s, there is now a fairly uniform deposition of fallout plutonium over most of the world, and especially over the northern hemisphere. Therefore, whether we realize it or not, we are living with plutonium every day, although fortunately in very small quantities. In some locations, however, environmental levels of radiation are significantly above background radiation; all are connected in one way or another with the weapons program. The environs of the Rocky Flats Plant near Denver, which fabricates warheads, is probably the location that has gained the greatest amount of public attention. What is important to realize is that there have been releases of plutonium in the past, and there undoubtedly will be releases in the future. The works of man are, unfortunately, not infallible.

From the viewpoint of risk, one must realize that most of the transuranium elements are alpha-emitters, which means that they must be inhaled or ingested and be retained in internal body organs to do any harm. The important parameters in a technical evaluation include defining the magnitude of the source term, the transport mechanism through the environment to human beings, the biological uptake and retention factors in human beings, and determination of a risk factor to the individual or to the population. The Environmental Protection Agency published, on November 30, 1977, a proposed *Federal Radiation Guidance on Dose Limits for Persons Exposed to Transuranium Elements in the General Environment*, which considered all these parameters in detail. (For more detailed information, see some of the background technical documentation published in support of this report.)

Our first consideration is the source term. Plutonium is, of course, produced in all uranium-fueled reactors and is itself a fuel—especially in the liquid metal fast breeder reactor. The primary concern then is with both the probability and the magnitude of a release. Examination of these factors has been done in a very comprehensive manner for light water reactors by the Nuclear Regulatory Commission under the direction of Dr. Rasmussen. Other components of the fuel cycle, primarily reprocessing plants, have not yet been subjected to a similarly exhaustive study. Neither have the components of the fast breeder reactor cycle. Certainly before we make a full-scale commitment either to recycle uranium fuel or to the breeder reactor, a study similar to that for the light water reactors would appear to be essential.

Environmental transport of the transuranium elements is a very complicated subject; the chemical and physical states over the very long time periods during which these elements may remain accessible to man must be considered. Air constitutes a pathway both for meteorological dispersion from the original source and for the resuspension of deposited particulates from the soil surface, both leading to possible intake by inhalation. Food and water provide a pathway for uptake by both plants and animals. Much research has already been done in this area, and more is in progress, but we have a record of only about thirty years with which to make predictions over perhaps hundreds or even thousands of years.

Next comes the question of relating quantities of inhaled or ingested radionuclides to dose to the individual. Dose rates to internal organs such as the lung, bone, and liver can be derived from models that relate ambient concentrations to absorbed dose. Such models are based on extensive animal and some human experience and are reasonably well established. Further refinement can be expected as more data are accumulated.

Finally, the dose absorbed by an organ must be related to the risk to an individual. For plutonium, such risks are both somatic in terms primarily of lung, bone, and liver cancer, and genetic in terms of risk to future generations. Derivation of such risk estimates is a sizable undertaking, involving extensive statistical studies on selected population groups. The risk estimates usually considered most appropriate for this purpose are those published in 1972 by the National Academy of Sciences' Advisory Committee on the Biological Effects of Ionizing Radiation. These estimate statistical risks for specific types of cancers in terms of total population dose.

Risk estimates can be presented in two forms, either the risk to an individual or the cumulative risk to an exposed population, often designated the "population dose commitment." Both are important in public health protection and in the decisionmaking process. The risk to an individual can be viewed in terms of absolute values; namely, it is either acceptable or not acceptable. This is especially true if the individual at risk can be identified. There are many parameters in judging risk acceptability, including whether the imposition is voluntary or involuntary, what the gain is to the person involved, and similar considerations. In practice, knowledge of the risk involved allows the individual to make an informed choice. In regulatory decisionmaking, the views of society as a whole must be considered and judgments made accordingly.

The population dose commitment is a statistical concept that adds the doses received by all individuals over a time period that can be as long as necessary, to derive a cumulative dose in terms of "person-rem." This cumulative dose can then be related to an expected excess number of incidences of cancer in the entire population. Such a calculation does not specifically identify the individual at risk, and is, in fact, based on models that generally group large segments of the population. This concept is of greatest use in cost-benefit analyses, which consider the risks and benefits of an activity to society as a whole.

It is obvious that the subject of risk and rationales of protection is one of the principal components of the process of decisionmaking concerning nuclear power. The judgments needed are complex and often controversial and depend on the availability of adequate technical information. The risks associated with transuranium elements in the environment are reasonably well understood and can be assessed on a quantitative basis, but the question of balancing them against their benefits to society may be much more difficult.

✳ *Chapter 7*

The Sociopolitical Point of View

Gene Rochlin

The presentation by Dr. Rasmussen (Chapter 5), addressing the question of defining criteria from a social perspective, is one of the most thoughtful pieces I have heard on this subject. It presents us clearly with the problem of what to do with an overall comparison or analysis of risk, once you have it. My comments will be addressed to the logical consequent step, "What do we have to do to turn these ideas into a form in which they can be applied and put into practice in a real-world social and political context?"

I will assume, and not without reason, that there is a difference between the "we" who are here, and I include myself as a physicist turned social scientist, as well as the formally trained social scientists who are present, and the "they" who are somewhere outside. In the Chicago area alone, there are several million of "them," and I suggest that there is a major difference between the way *we* approach this problem and the way *they* perceive it. One of the reasons has to do with the public's perception that science has "sinned," not in the way that this notion is often stated in critical or dissenting scientific literature, but in a more mundane way.

Surveys have noted that scientists continue to command greater public respect than engineers, largely because their work is considered to contribute to society rather than to manipulate it. But in some sense, the scientific community is perceived to have "sinned" by becoming an interested party in the particular debate that concerns us in this symposium. The remoteness from decision that might once have been stipulated as a scientist's distance from the issues, the scientist's ability to analyze impartially without seeking to gain from the outcome, has been badly compromised.

The failure of groups such as this one to come to terms with such perceptions is illuminated by the statement at the beginning of Dr. Rasmussen's paper to the effect that societies have always sought technologies and progress. That is not exactly correct. Individuals and small groups in societies have sought to promote and exploit new technologies—generally to advance their own power and social position. Development of new technologies is often perceived by societies as being to the general benefit, and undoubtedly it often is, but it is not for that reason that most technologies have been developed. A basic premise of engineering education in many schools is that the engineer is also an entrepreneur. Technologies are sought and promoted for individual gain. Let me hasten to point out that this is a description, not a criticism. The number of individuals who govern their actions by what they hold to be best for society rather than for themselves is small. What is more, I usually do not trust them.

Given the basic premise that engineers are one among many groups seeking position and power within society, we may ask what do "they" do when they feel their voice is not being heard? The controversy over nuclear waste disposal is a good example. A concern has been voiced that even if there are safe technical solutions to the waste disposal problem, only part of the problem is being addressed. There is concern that the agencies in charge will not carry the operations out properly and safely. The political responses manifest themselves in many ways, not the least of which has been the passage of NEPA or the Delaney amendment. These responses occurred because large groups in society believed they had to institutionalize their right to be protected. Much of the battle between administrative agencies and the Congress represents a fight between vested interests, who are much better represented by lobbyists, and those who believe they are created to speak for the nonvested, who are unable to bring specific political pressure to bear.

Governmental agencies often have to apply blanket rules, because to make exceptions appears to be a yielding to special pleading. For example, drugs that may be relatively useless are subjected to lengthy and very tight testing criteria because of potential risk, but so are drugs that could have enormous benefits. It is too difficult for an administrative agency to discriminate yet remain credible.

A second difficulty in making use of or acting upon risk analysis is human fallibility. Reading over the sequence of events that led to the 1977 Consolidated Edison blackout in New York, I was struck by the fact that it took several different actions by individuals doing the wrong thing, or the right thing at the wrong time, to bring on the collective collapse of the system. In a sort of inverted causal fault tree, if any of these people had failed to take the incorrect action at the proper moment, the blackout would have failed to occur.

In the case of the San Diego air crash of 1977, the new automatic collision warning system had been installed only a few weeks before, and even though the alarm went off, there was no command for emergency evasion made to the pilot by the control tower. More to the point, perhaps, is the ruthenium blow-back accident at the Windscale plant in the United Kingdom in 1973, when the first person to observe the radiation monitors responding thought the system had malfunctioned, because so much ruthenium had been released that all of the individual monitors had gone to full scale. Had that person's first conclusion not been that there was a systemic failure, evacuation would have proceeded more rapidly and worker exposure would have been reduced.

It is difficult to factor such events into a risk analysis, although some examples are included among the possible faults in such advanced documents as WASH–1400. The public perception, however, remains that quantifiable risk analysis does not tell you everything you need to know.

Returning to the question of plutonium risk, there is another social issue that does not appear to have received much notice. There has been a shift in attitude from that of the 1950s, which might be characterized as: "Do we love plutonium fuel cycles?" to that of the late sixties and early seventies, which was toned down to: "Do we want plutonium fuel cycles?" Now, even those here who most strongly support the use of plutonium as fuel are taking another tack and arguing that we must have plutonium fuel cycles because there are no available midterm alternatives. In terms of the central perception of risk and tolerance of it, there is a big difference between what risk will be assumed for something that is wanted, and what risk will be tolerated for that which you are told you must do. These differences in attitude need to be factored in.

Plutonium use has become a test case, I believe, for many things that are going on in society. It is "on the point." It is not being judged "fairly" compared to other technologies and industries. But no one ever claimed that social and political processes were, or had to be, fair—let alone just or equitable. That is an ideal but rarely achieved in practice. Moreover, attempts to provide comparisons of risk levels to balance judgments can have paradoxical results.

An executive of a major nuclear company once pointed out that the potential risk from nuclear waste is no greater than that from chlorinated hydrocarbon pesticide wastes that are being disposed of. My response was: "Fine. How would you like to put that in your company's advertising?" Looked at that way, the matter of comparative risk is less simple to deal with. To take another example, one California city had to reexamine the chlorination of its water because its residents had suddenly discovered that there are liquid chlorine tanks within the city limits. Perceptions shift. To say that shipment of nuclear material is no more dangerous than that of chlorine is of little use if the public reaction to the awareness of the dangers of chlorine shipment triggers a call for tighter controls.

The social relativity of risks is difficult to establish. There is, of course, the famous statistical analysis of Starr that shows that there are differences of acceptability between voluntary and involuntary risks. Clearly, at some level there is a perception of excess risk from some activity, and society intervenes, as with seat belts. But most studies such as Starr's are post hoc and macroscopic. They are not the type of studies that are required for an understanding of individual and collective responses to and perceptions of future risks, or what cost society will bear to reduce them.

One of my favorite examples of social relativity is that of kidney dialysis machines because there are not that many, and these are lives that can be visibly saved. There are other places in medicine where that same money, if well spent, could save more lives either in the present or in the future. But dialysis is affordable, effective, and highly visible. In a very human way, we have chosen to do what we can where we can, and not respond to a more systematic cost-benefit that would sacrifice those lives openly, for the sake of saving others that are less visibly affected.

Yet another social aspect of risk assumption that is not much discussed is that of commitment. There is a tendency to take a higher risk for something for which resources are already committed than for something not yet started. For example, look at the Diablo Canyon nuclear power plant in California. It is built, and Pacific Gas and Electric has more than a $1 billion investment in it. To write that off and not open the plant would require that each of the company's customers assume a burden of several hundred dollars to pay off the capital costs and interest. Pacific Gas and Electric would not like to do that, and has fought to get the plant licensed. If it were still in the planning stage, I believe that they would have written off the site rather than take the risk of public protest. Moreover, I believe the public would have been more reluctant to accept the site as a proposed site than as a site on which a plant has already been built.

To return to our most comprehensive example of risk analysis, consider the loss-of-coolant accident (LOCA). I do not intend to get involved in the question of how many people could be killed, or what the future cancer incidence might be in the event of a LOCA. Let us assume that there is a major meltdown, with complete loss of the plant but few immediate deaths at the site. This will be perceived by the public as a major catastrophe. The cleanup will be expensive and difficult, and both the accident and the following response will be front page news all over the world. My opinion is that such a sequence will become more acceptable as the societal commitment to nuclear power increases. If a LOCA had happened five years ago, it might have crippled the nuclear industry. If it happens within the next five years, it would probably result in a major delay in deployment. If it should not happen until the time when perhaps half of our total electricity is nuclear-generated, there will be tendency to say, "That is the price we have to pay."

There is some realization that a commitment entails the acceptance not only of known risks, but of future risks that are uncertain, so that you may get stuck with unfortunate consequences if risks turn out to be higher than believed. Although I hate to indulge further in what my colleagues often term "anecdotal social science," there is some evidence that those who enter into a commitment (such as marriage) are willing to act on the basis of less perfect information if they feel there is a way to reduce or remove either the commitment or the risk in the future.

Finally, there is the matter of more general, and therefore less easily capturable, psychological social response. In the group I work with at Berkeley, the consequences of the use of a technology are characterized as first, second, or third order according to the directness of coupling between the technical stimulus and the social and political adjustments. First-order consequences are direct: you use something, get benefits, and incur costs. That is relatively easy to measure. Second-order costs are those incurred in order to accommodate to the first-order ones. Creation of the EPA is a second-order cost; it meant putting a new bureaucracy in place with certain types of delay, and social and political conflict engendered by the existence of the new agency.

The most subtle effects are the third-order ones—alterations in individual and social perceptions of a society and the world in which it is embedded. When a technology is deployed, the direct benefits and costs occur, followed by the second-order social and political adaptations to use, regulate, and control both the technology and its direct effects. But the final outcome is a change in general views as to what the world is like. The world we perceive, given the existence of a widespread and effective telephone system, for example, is quite different from that perceived before that mode of communication existed. This is not just a matter of social, political, or technical analysis. It is a *different* world, and both individuals and societies act quite differently than they would, or could, otherwise.

These are the more important underlying perceptions, clearly related to the question of hard versus soft energy paths as well as to the use of plutonium. There is some uneasiness about what is being committed by the choice, whether the more general one of hard versus soft paths or the more specific one of utilizing plutonium fuel cycles. More and more people are beginning to appreciate the depth and breadth of the changes in society that will follow upon the choice of energy supply we make, the understanding that the world that results will not be the same as the one we have known. Much of this usually unfocused and intuitive concern appears to have coalesced around the plutonium decision so that the use of plutonium will be scrutinized to a degree unmatched by any other technological decision to date.

The matter of whether this is fair or just is largely irrelevant. More to the point is to try and determine to what extent the kind of analysis we have been discussing thus far will contribute to the outcome. I believe that the type of general definition of criteria that Dr. Rasmussen has put forth will be both reasonable and useful. At the very least, it begins to deal with the more general social and political perceptions that underlie the present debate. When "we" start to throw up a screen of numbers, and in particular, when "they" observe us to quarrel as to the precision of the numbers and their applicability, we only increase further the general uneasiness that has been the catalyst for the coalescence of a wider range of social and political concerns about the use of technology onto this specific issue.

Discussion

V. Paul Kenney: Who is the "they" we're discussing? Doesn't it seem that a sizeable component of this "they" is just opposed to the very idea of nuclear energy? "They" argued that reactors were apt to "go off" like atomic bombs, and the nuclear engineers patiently responded by explaining that they could not go off like a bomb. "They" then expressed opposition because of the radiation given off by reactors in normal operation, and the nuclear engineers responded, "It isn't much; coal gives off more." "They" argued that what "they" were really worried about was the thermal pollution problem, and the nuclear engineers responded, "All power plants give off heat." Now "they" argue that what "they" are really opposed to in nuclear energy is the waste problem, the plutonium problem, the proliferation problem, the problems we are addressing in this symposium.

I wonder if it is not just that "they" are not so much opposed to nuclear energy because of these various things, as that "they" bring up these things because they are so opposed to nuclear energy itself. There is an identification in many people's minds between nuclear energy and Hiroshima and a general guilt. There are a large number of people who are just against nuclear energy. No matter how patiently the nuclear engineers respond, they are just going to raise an endless succession of further objections.

Gene Rochlin: Take a close look at the extant public opinion polls on nuclear energy. The first thing that you will notice if you look carefully at the questions is that, depending on the questions asked, three-fourths of the public approve of nuclear energy or two-thirds oppose it. That is because the questions asked do not capture things very well. They tend to be rather unsubtle. Correlations to other concerns are rarely sought. If you polled high school students today, four out of ten would probably think Hiroshima was a Japanese beer.

I do think there is substantive unease about nuclear power and plutonium or about the kind of social system it takes to support it. Plutonium use looks worse because it looks very complicated; there are so many potential problems, and so many things you have to do to prevent them. I think that this is a major source of public unease (and here I am prophesying something we have yet to test); to the extent that public concern is ultimately deep-seated, it tends to attach to the prominent problem. Thus, if reactor safety concern drops, and public concern shifts to waste management, it is because the underlying concerns do not derive from the *specific* technical points. They are much more general. One does not see any survey or public opinion poll today that really examines this very much. To be fair, it is probably a difficult idea to catch by public opinion polls at all.

Herman Daly: In his very thoughtful remarks, Dr. Rasmussen suggested, if I understood him correctly, that our criterion for use of plutonium should be that the use in question not make it easier to make a bomb than is currently the case by concentration of natural uranium. Does this imply a freeze on the current technology of uranium concentration, so that we have a fixed reference as a criterion, or do you see the technology of uranium concentration as continually improving, making it ever easier to build a bomb, so that we would have an upward-sliding benchmark?

Norman Rasmussen: I am sure that isotope separation technology will continue to develop and become easier. Although this technology has been classified as secret by those nations that possess it, it is inevitable that it will become more widely available. For these reasons, I believe it will become easier for nations that want nuclear weapons to obtain them. In other words, the technical barriers will get lower and it will be more important to develop political incentives for not building nuclear weapons.

Harvey Brooks: I'd like to hear Norm Rasmussen's comment on one point that I think would be worth bringing out in connection with the risk criteria. One could argue that as more reactors are deployed there is a case for saying that the probability of accidents ought to become smaller, that is, the unit probability, in order to keep the overall probability either constant or decreasing.

This is certainly what happened in the case of aircraft, and, in fact, there is a very interesting paper by B. Lundberg, written in 1962, in which he extrapolated the then existing statistics of aircraft accidents, and showed that by the year 1979, there would be one accident per day somewhere in the world that would kill between 100 and 200 people. He used that as a very strong argument to the aircraft industry to get busy, or "they'll shut you down." The industry took that very seriously. I wonder. When I have said similar things about nuclear safety, many people in the industry say, "No, we should adopt a certain criterion of safety, and that is for all time."

Dr. Rasmussen: I am perfectly willing to accept that we should have a criterion for all time, but it should be at the specified risk level; as we get more plants, the safety of each plant has got to be better. I implied in what I said today that whenever something new is introduced, it has got to be better. It cannot be introduced at a higher risk level. I could not accept a situation where there is one core-melt accident every two or three years. If we pay attention to the business, we should learn from experience. Thus, it should be possible to reduce the risks from individual plants as effectively as the aircraft industry has reduced risk.

Lester Lave: I like the argument that Dr. Rasmussen was making, but I do not think that the conclusion follows from the argument. One of the things you said (a point I think is very well taken) is that it does not make any difference to set a safety standard unless you agree on how to measure it. Lowering the risk to 10^{-7} does not mean anything unless you agree on how to establish this level. The example that you used was the Dutch example of one North Sea storm in every 10,000 years for dikes. After all, 10,000 years ago, the ice age was receding; the level of the North Sea was a lot lower and thus the chances of flooding lower. Ten thousand years from now, if the effect of the CO_2 increase is as expected, the Netherlands may be completely under water, no matter how high they build their dikes.

Dr. Rasmussen: No, you should not think of it that way. It is a risk of one in 10,000 per year that is the criterion, and not one in 10,000 years. There is a big difference between the two.

Dr. Lave: You also mentioned, a couple of times, the alleged lack of comparison of the health effects of coal mining with nuclear, or of coal with uranium mining.

Dr. Rasmussen: I did not say that.

Dr. Lave: I hasten to point out that there have been a number of studies of that sort, and as far as I know, they have all come out with the same sort of conclusions. Aside from the risks that are not quantified, light water reactors are anywhere between 100 and 10,000 times safer than coal.

Dr. Rasmussen: Yes, I am aware of that and I hope I did not imply that health comparisons were not possible. They are, but to compare climate modification with proliferation is, I feel, very bad.

Dr. Lave: Anyway, because of such problems, it is difficult to assess the risks, more difficult even than the very clever criteria you proposed can handle; for example, plutonium versus uranium isotope separation. Inevitably, real-world distributions have fat tails, instead of the nice tails in theoretical distributions. And, inevitably, the theoretical probabilities that one calculates understate what one actually finds in experience.

There is much psychological evidence that people, in trying to estimate real-world probabilities, vastly overstate their confidence levels; they see many fewer possibilities and overestimate the occurrence of familiar events. These findings tell us something about why people tend to be suspicious of scientists in general (not just in their statements about nuclear power). That suspicion is well founded.

There are some nice studies that were done by a group out at the Oregon Research Laboratory (Fischoff et al), who have been trying to discover the dimensions that people use to evaluate acceptable risks. They provide some generalizations of the notion that Chauncey Starr proposed in *Science* some years ago. He wrote about the importance of an action being voluntary rather than involuntary. Fischoff has found important dimensions to be the familiarity of the action and consequences. I think that these findings are instructive; at least they provide me with insights as to why nuclear power has generated so much controversy, compared with, for example, automobile driving. Clearly, automotive risks are much greater than nuclear risks, but people are less worried about them.

I conclude that it is really inherently impossible to get a very good handle on how safe technologies are; there are always aspects which are impossible to quantify. If we begin to accept such an observation, we are left, inevitably, with the conclusion that (at our current levels of real income) society will desire to slow the rate of introduction of new technology. No matter how much analysis is done in advance, society wants extensive experience at each stage before advancing to the next and becoming utterly committed to a technology. We can look forward to much slower introduction of technology in the future, no matter how good our analyses of safety are.

Dr. Rasmussen: I guess that is Lester's opinion, not really a question.

Dr. Rochlin: I just want to make a small comment about the automobile. What surveys and tests often show is that people do not believe that their own automobiles will have accidents, even though automobiles in general have a high accident probability. One of the best examples is people who use their seat belts on the highway where they feel that other people can hit them and do not use them for short trips to the supermarket, even though, if you look at the statistical tables, the probability of accident is just the reverse. You should belt up to go around the corner and not worry so much on the highway.

Bernie Cohen: As soon as you say that nuclear power is much safer than the methods that we are using now to get electricity, that is the end of the story. We do not need all these criteria and all this philosophy. Even the Union of Concerned Scientists agrees that nuclear power is much safer than our other methods of getting power. We only have one problem: it is to make the public understand that simple fact. The source of the problem, it seems to me, is not all that psychology, sociology, and so on; you have to stop people like Robert Redford from going on television and lying to the public. That's where the problem is!

Alvin Weinberg: Let me say that I thought Dr. Rochlin's observations on what Dr. Rasmussen said were really very much to the point. I was delighted that Dr. Rasmussen was, in effect, putting forward what I hoped was a very important consideration for this community, a new way of looking at the whole question of establishing standards. In essence, Dr. Rasmussen offered a response to what might be called the absolute metaphysic, which Dr. Daly proposed. You are proposing that we adopt a kind of relational metaphysic in trying to establish what are acceptable risks. I can only say that I think that is the right way to go after it.

Twenty years ago, the first Friedell Committee report, which examined the effect of low-level radiation, said that the sensible way to answer the question, "What is the acceptable risk for low-level radiation?" was not to start with the occupational standards, which were entirely arbitrary, and move downward, but to start with the background and move upward. In twenty years we have strayed farther and farther from this sensible suggestion. However, I was told earlier by Dr. Burley that the standards that the EPA now establishes for the entire fuel cycle, do, in a de facto way, get back to the original proposal of twenty years ago. The standard is 25 mr per year for the entire fuel cycle, which happens to be, as Dr. Burley pointed out, just the standard deviation of the natural background. It seems to me that the ideas that you have cast before us are ones we ought to think about very seriously, and the regulatory bodies also ought to think of them seriously. There are many possibilities for setting standards of exposure at a value comparable to what we have become used to and find acceptable.

Now, with respect to my good friend, Dr. Cohen, who points out that nuclear power is marvelously safe, I agree. "Then," you can ask, "how in a weak moment did Alvin Weinberg invoke Faust?" Many people, including myself have asked me that, so I turn around and put the question back to you, Dr. Rasmussen. I am glad that Dr. Zivi is in the audience, because he has thought about these things, and was the first to bring them to my attention. Namely, when you talk about the risk from nuclear power, if you look at the risk measured by linear average, the mean probability multiplied by the consequences, the first moment of the distribution, then what Dr. Cohen said is absolutely correct. I do not really think that is the issue.

I was at the Lemoniz reactor, in the Basque country of Spain about three weeks ago, and what the people there are concerned about is not the fact that the average probability is very small; they are concerned about whether the city of Bilbao will possibly have to be evacuated, let alone the question of whether people really get killed. That, I think, is the issue, and I would like you to say a word about that.

Dr. Rasmussen: I think that has to be an issue; I do not know what I can say, except, sure, anybody who is near a plant worries about what it might do to them. I went to the Barseback plant; the Danes were worried because something might happen to cause the evacuation of Copenhagen. You can see the city of Copenhagen sixteen miles away; so it has to be a concern. What you are asking is, "Is there some unusual set of events that will make it much more likely than the average, to happen here, in this place?" Is that the thrust of what you mean?

Dr. Weinberg: Effectively, in spite of what Dr. Cohen has said, we are talking about a real issue, and the real issue is that people in their perception of this whole business are not looking at the average, they are looking at the unlikely, very large incident.

Dr. Rasmussen: Well, there is no question that they are.

Dr. Weinberg: Therefore, finally, whether we can accept it or not, it is a political decision.

Dr. Rasmussen: I do not quite agree with Dr. Cohen on one thing. I think you can compare health effects, which have roughly the same consequence, but when you talk about climate modification and proliferation, you cannot measure them on the same scale that you measure health effects. You have a much tougher comparison when you ask about the whole fuel cycle than just the effect of radiation or the public health aspect of that. That is why I think it is such a difficult problem. You cannot just say that we are safer than with coal, because coal does not really have a proliferation problem; but it does have a climate modification problem, and I don't know how to weigh those two on the same balance. That's what troubles me about your position. I have read Dr. Lave's article about what the relative health risk is, and I believe it represents a reasonable comparison of health risks, but that is just one consequence. It does not include proliferation and it does not include some of the other things.

Stephen Beckerman: I'm quite impressed with Dr. Rasmussen's ingenuity in taking these different scales of tolerable safety and relating them to one another, but, at the same time, I find myself uneasy, and I suggest the acceptability to all of us here of this impressive ingenuity has to do with the fact that we are all academically trained, and we have been trained to appreciate this sort of thing. I believe that if the "they" that Dr. Rochlin talks about were to be confronted with this sort of reduction of everything to the same scale, "they" would find it not only unacceptable, but offensive. To a certain extent, I agree with that myself. I believe that we can demonstrate fairly easily that people do not have a unidimensional value system for evaluating events in their lives, but that they have many value axes, and that these axes are in some sort of hyperspace, orthogonal to each other. To attempt to project all the values that move along different dimensions onto a single scale is to do a certain violence to the way that people conceive of their world. I doubt very much if anyone here could live his life in accordance with the sort of comparative evaluation you have proposed. (Dr. Rochlin gave an excellent example of the seatbelts in a car.)

Dr. Rasmussen: I do not know what you mean by the same scale. Maybe my final paragraph there implies that a death is a death and we ought to weigh them equally by having a uniform agency whose goals were to save the most lives per dollar invested? Is that the issue? Because I have not the same scale in trying to cope with waste, and how to cope with proliferation?

Dr. Beckerman: You say, "Reduce the risk of atomic proliferation by making it more difficult to make a bomb from plutonium than it is now to make a bomb from uranium." I think that is extremely clever, and I like the idea; yet, on the other hand, it does a certain violence to a schema in my head about the different dimensions along which these things move.

Dr. Rochlin: There was an informal survey done on that among some students at the University of California. What very often happens is that when you tell people that "A," which they believe to be dangerous, is less dangerous than "B," which they already accept, they do not accept "A," but start rejecting "B"!

Marc Roberts: The discussion we've been having leads to one of the key issues of the whole safety area. Are we willing to accept the decision theoretic model for thinking about nuclear accidents? That is, do you use, as Dr. Weinberg pointed out, the first moment of your probability distribution over the probabilities, or various events? Decision theory says, "Yes, you just take your best guess as to these probabilities." Insofar as you have attitudes towards risk, those attitudes don't show up by monkeying with the possibilities. Rather, they show up in terms of the values you place on alternative consequences. Risk aversion comes through having a particular kind of nonlinear shape to one's utility function. It has been striking, I think, that in the study of nuclear safety, trained academics, when first confronted with the austere logic of decision theory, have often rejected it. People instead say, "I don't want to use the first moment of my probability distribution over the probabilities. I want to use the upper bound of my probability distribution over the probabilities, as the accident probability for the analysis."

The only claim that decision theory ultimately has on us is as a consistency condition. It says that if we make decisions any other way, our choices among alternatives will not necessarily be transitive. It is an implausible implication, that transitive choices require one to use the first moment of this distribution of accident probabilities. It seems to me part of the problem in getting people to do that is that they do not then go on to correctly specify their utility functions. Dr. Rasmussen may have done just that by looking at everything in terms of expected lives lost. You have assumed, in effect, that the utility function is linear in lives, so that one-half the probability of losing 1000 lives is the same as the certainty of losing 500 lives, and so on. It may very well be, I suggest, that for many of us, an event in which we lose 1000 lives is more than 1000 times worse than an event in which we lose one life.

Dr. Rasmussen: That's for sure true.

Dr. Roberts: But if you do that, then you cannot just look at expected lives lost.

Dr. Rasmussen: Absolutely, but where did I ever propose that?

Dr. Roberts: It seems to me that, like many people who have talked about whether nuclear power is safer than coal, you used an expected life lost metric. If we differentially weight the lives lost, on the basis of the number of lives lost in an event, we could change the answer somewhat. That might be truer to the spirit of decision theory, which implies the possible use of a nonlinear utility function.

Dr. Rasmussen: Yes, I think we are risk averse to big events; no question about it. But if you look at Dr. Lave's work, and so forth, most of the risk came from very similar kinds of events. There were long-term cancers from low-level radiation, versus long-term respiratory disease from low-level coal; it is perfectly fair then to compare them on a one-to-one basis. In my general suggestion at the end, that somebody allocate money versus a life, it should not be a dollar per life directly, because you have to take into account his risk aversion to big events. I think society is generally that way.

A.M. Saperstein: I want to respond to Dr. Brooks' comment and your response. He said that the probability of malfunction had to decrease as the number of units increased, and your response was "Naturally, experience will lead us to improvements."

Dr. Rasmussen: Only if we pay attention and take advantage of it.

Dr. Saperstein: Of course, our history has been just that; experience has led us to improvements. But the experience almost invariably has been an experience with accidents. We've had boilers blowing up, cars crashing, airplanes falling, and so forth.

Dr. Rasmussen: We've had 10,000 failures in nuclear plants that we've learned an awful lot from.

Dr. Saperstein: I expect the general public would consider all of those as minor failures, whereas an airplane crash . . .

Dr. Rasmussen: As to the safety situation, these failures provide just the information we need in order to keep the system from failing or to reduce the probability of failure. They are minor accidents but, coupled with other things, they could have been major accidents. Obviously, it is the series of minor accidents that led to a major accident. So all our experience in small events is very useful in lowering the risk rate of large events, if we take advantage of it.

Dr. Saperstein: The point I'm trying to get to, speaking for myself as "they" for the moment because I am most of the time, is that the public is not finding it easy to accept a small accident in the same category as a major accident—saying you can learn the same thing from both of them. Perhaps that is a psychological hangup, but it is something you have to deal with.

Dr. Rasmussen: Not as long as we do not have major accidents.

Dr. Saperstein: You do, if you want to persuade people.

Dr. Rasmussen: Well maybe, but we are getting along fine. We are convincing them with our record. Four hundred or so plant-years without a major accident is a pretty convincing record. So I think that if we repeat that for another decade as someone said, we will be into the nuclear age to the point where people will accept it for the reasons Dr. Rochlin pointed out.

Dr. Brooks: I wish to remark on Dr. Roberts' comment that there is anything inherent in decision theory that requires using expectation values. You can weight the distribution theory that requires using expectation values. You can weight the distributions of accidents any way you please; it does not change the analysis; it changes the quantitative outcome, but it can be fed in, in any way. It seems to me it will be useful to put in various degrees of risk averseness into the distributions of accidents. I would point out, as Dr. Rasmussen hinted, but I think is not sufficiently appreciated, that most of the deaths that occur from nuclear accidents that one has in your analysis are very similar to the coal deaths. That is, they are cancers thirty years down the road, due to exposure of large populations to very low-level radiation. In fact, the thing is much more comparable, I think, than much of the discussion on the subject tends to indicate. The number of deaths of this kind usually tends to be, in most cases, twenty-five, thirty, or even more times the number of prompt deaths; in fact, for most accidents there are no prompt deaths, only deaths due to low-level radiation that are long delayed in time.

✳ *Chapter 8*

Some Comments on the Toxicity of Plutonium-239 [a]

R.E. Rowland

Not too long ago the toxicity of plutonium was a subject of interest only to a few radiobiologists. Today, however, profound comments on this subject can be seen in daily newspapers and weekly news magazines and heard on radio talk shows and television news programs. By such news-media comments the views and opinions of the general public are obviously influenced. Indeed, since the media thrives on catastrophes, hazards, and establishment "cover-ups," and since there is little news value in lack of catastrophes, hazards, and cover-ups, it is to be expected that the picture presented of plutonium toxicity will be grim.

Plutonium is a very toxic material. But it is no more hazardous than many other elements and compounds we utilize in our complex world. The purpose of this paper is to attempt to provide a rational framework for the evaluation of some of the hazards of plutonium and to compare, numerically, those hazards with some of the risks inherent in everyday life.

A great deal of concern about the toxicity of radioactive materials arises from the assumption that the effects are linearly related to dose. In fact, very little information exists about dose and effects in human beings; what information is available applies only to very high doses. There is no doubt that, when little or no information is available, very conservative guidelines should be utilized. The linear hypothesis is such a guideline. However, such guidelines have no validity as predictors of effects at low doses, in spite of the fact that they are continually misused in that fashion.

[a]Work performed under the auspices of the U.S. Department of Energy.

This presentation will suggest criteria for the evaluation of two aspects of plutonium toxicity related to the known ability of alpha-emitting radionuclides to induce malignancies in human bones and lungs when deposited in those tissues. A method for obtaining a quantitative estimate of the risk of the induction of bone cancers will be outlined, based only on data from human beings, and a qualitative examination of the lung cancer risk will be attempted.

Some quantitative estimates of risk will be presented that may suggest that we tend, perhaps intuitively, to have a concept of "acceptable risk." Whether one knows the magnitude of the risk or not, there seems to be a general understanding of those activities that are considered risky, and those that are looked upon as relatively safe.

What is the risk of death that faces us as members of a well-developed civilization? As a crude approximation, one can state that we live with a 1 percent chance of dying within the next year. This is based on the fact that, out of a population somewhat greater than 200 million, almost two million people die each year. We recognize, however, that the probability of death is a strong function of age. Table 8–1, which gives death rates for 1973 as a function of age, illustrates this strong dependence. Most of us would agree that any additional risk as large as the smallest value given in this table, the value of forty-one deaths per year per 100,000 people, would be considered excessive.

Table 8–1. U.S. Death Rates for 1973

| Age Group | Deaths per 100,000 Population per Year | | Ratio: Cancer/Total |
	All Causes	Cancer Only	
5–14 years	41	5.4	0.13
15–24 years	128	7.5	0.06
25–34 years	154	15.6	0.10
35–44 years	296	56.1	0.19
45–54 years	697	180.9	0.26
55–64 years	1612	430.0	0.27
65–74 years	3440	768.1	0.22
75–84 years	7932	1187.9	0.15
Over 85 years	17429	1435.3	0.08
Overall	940	167.3	0.18

From *Vital Statistics of the United States* (1973) *Volume II—Mortality*, Part A.

What kinds of risk face us in our various occupations? Table 8−2 lists the incremental risks for some occupations as published by the U.S. National Safety Council for 1975. It is evident that the suggested value of forty-one deaths per year per 100,000 people as an excessive incremental risk does fall between the rate of risk for those industries one might consider safe and the rate for those considered to carry an element of excessive risk. The overall average for all industries, fifteen deaths per year per 100,000 employees, can be used as a reference level. Any occupation with a value higher than this is more dangerous than the average.

THE MANCUSO STUDY

The effects of low levels of radiation on exposed individuals has been a topic of considerable interest for some time, especially since an evaluation of the number of malignancies that have appeared in the workers at the Hanford Works at Richland, Washington, was publicized.

Of four reports on the Hanford workers that I will mention, two indicate that there is an excess of cancer among these workers due to their exposure to ionizing radiation and two have contradictory conclusions. Each report by itself seems convincing and thus raises questions. How does one decide between them? What is the truth?

Each report uses essentially the same population, or at least a group derived from the total population. Each report makes use of the accumulated exposure as determined by film badge readings recorded for each worker. An essential factor about these accumulated exposures must be kept in mind; the longer an individual worked at Hanford, the more opportunity the individual had to accumulate a

Table 8−2. U.S. Work Accidents, 1975

Industry	Deaths per Year per 100,000 Employees
Trade	6
Manufacturing	8
Service and government	10
Transportation and public utilities	33
Agriculture	58
Construction	61
Mining and quarrying	63
All industries	15

From the National Safety Council Accident Facts (1976).

higher reading; and the longer an individual worked, the older that individual became. Therefore, there is a correlation between age and dose.

Likewise, one must keep in mind that in the United States today about one out of every five deaths is a death with cancer. Actually, this ratio increases from about one out of ten for age thirty to one out of four for age sixty, then decreases with advancing age, as shown in Table 8—1.

Mancuso, Stewart, and Kneale (1977) concluded that incidences of bone marrow cancers, pancreatic cancers, and lung cancers were elevated among the Hanford workers. Their conclusions were derived from the observation that those who died with cancer had accumulated higher external radiation exposures than those who died without cancer. This effect is indicated in Table 8—3.

In a subsequent report the same authors concluded that approximately 5 percent of the cancer deaths of Hanford workers were radiation-induced and that these extra deaths were probably concentrated among cancers of the bone marrow, lung, and pancreas (Kneale, Stewart, and Mancuso 1978). Their data, given in Table 8—4, illustrate that the proportion of cancer deaths increases with increasing exposure.

Marks, Gilbert, and Breitenstein (1978) concluded that multiple myeloma and carcinoma of the pancreas were elevated in Hanford workers, but since the forms of cancers commonly associated with radiation exposure, such as myeloid leukemia, were not elevated, the role of radiation as the cause of the cancer elevation was not proven. Their results, which included a comparison of the number of observed cancers with the number expected from U.S. statistics, are summarized in Table 8—5.

Finally, a report by Sanders (1978) concluded that the observed increase in cancer deaths among the exposed Hanford workers was not related to their radiation exposure. Table 8—6, extracted from this report, shows that the dose accumulated by all workers increased with time, and that while those dying with cancer did have larger accumulated doses than those dying without cancer, there was no

Table 8—3. External Radiation Records for Two Groups of Nonsurvivors: Cancer and Noncancer Deaths

Nonsurvivors	Number of Deceased Workers	Mean Radiation Dose (rads)
Cancers	670	1.38
Noncancers	2850	0.99

Extracted from Table 2, Mancuso, Stewart, and Kneale (1977).

Table 8–4. Fraction of Total Deaths Due to Cancer as a Function
of External Dose

Dose Level (rads)	Cancer Deaths	Noncancer Deaths	Cancer Deaths / Total Deaths
Less than 0.08	256	1068	0.193
0.08–0.31	131	592	0.181
0.32–0.63	119	428	0.218
0.64–1.27	123	448	0.215
1.28–2.55	91	320	0.221
2.56–5.11	48	147	0.246
More than 5.11	64	198	0.244

Extracted from Table 3, Kneale, Stewart, and Mancuso (1978); the last column
was calculated from the given data but did not appear in this reference.

Table 8–5. Observed Deaths, Expected Deaths, and Standardized Mortality
Ratios (SMR's) for Selected Cancer Types in a Selected Group of White Males
Who Worked at Least Two Years at Hanford

Cause of Death	Observed	Expected	SMR
All malignant neoplasms	414	487.7	85
Cancer of pancreas	28	28.1	100
Malignant myeloma and other cancers of hematopoietic tissue	33	31.5	105
Leukemia	10	21.6	46
Cancer of bone	2	3.1	65
Cancer of lung	115	147.7	78

Extracted from Table 2, Marks, Gilbert, and Breitenstein (1978).

increase in cancer with higher doses. This effect is also shown in
Table 8–7, where only those who were hired in 1944 and 1945 are
listed. The statistics for this group, which accounted for about 60
percent of the cancer deaths, show clearly that those dying with
cancer had higher doses than those without cancer, but also show
that a fourfold increase in average dose did not increase the proportion of cancer deaths.

It appears that, for some reason, there are more cancer deaths
among exposed Hanford workers than one might expect. However,
the fact that there is not an increase in cancer rate with increasing

Table 8-6. Comparison of Proportion of Deaths with Cancer, and Accumulated Lifetime Doses, Among Exposed Male Hanford Employees. Deaths for 1944-1972 for Employees Starting Work 1944-1971 *(total 17,600 employees)*

Years	Number of Deaths		Cancer Deaths (%)	Mean Cumulative Exposure of Those Dying (mrem):	
	Cancer	Others		With Cancer	Without Cancer
1944-1959	104	420	19.85	548	579
1960-1972	331	1302	20.27	2539	1963

Extracted from Table 1, Sanders (1978).

Table 8-7. Comparison of Proportion of Deaths with Cancer, and Accumulated Lifetime Doses, Among Exposed Male Hanford Employees. Deaths for 1944-1972 for Production Employees Starting Work in 1944 and 1945 *(total number unstated, but of the order of 6000)*

Years	Number of Deaths		Cancer Deaths (%)	Mean Cumulative Exposure of Those Dying (mrem):	
	Cancer	Others		With Cancer	Without Cancer
1944-1959	74	284	20.67	551	652
1960-1972	190	769	19.81	2282	1604

Extracted from Table 3, Sanders (1978).

Table 8-8. Summary of Japanese Atomic Bomb Experience

Number of survivors in 1950	285,000
Normally occurring deaths	70,000
Excess cancer deaths attributable to radiation	337-492

Extracted from Moriyama (1978).

radiation exposure makes it very unlikely that this increase is due to radiation.

Perhaps the difficulties of understanding the results of these studies of the Hanford workers are best illustrated by a review of the findings for the atomic bomb survivors in Japan (Moriyama 1978). Table 8-8 summarizes the population studied, which was about ten times as large as the Hanford group. Less than one-half of 1 percent of all deaths among the survivors could be attributed to radiation-induced cancer. Figure 8-1 illustrates the increased risk for certain

Figure 8−1. Relative Risk, with 80 Percent Confidence Levels, for Selected Causes of Death for Atomic Bomb Survivors, 1950−1972. This is a comparison of the risk for those who received doses greater than 200 rad with those who received 0−9 rad. Extracted from Moriyama (1978).

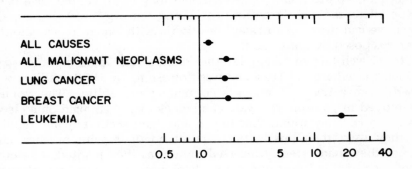

specific causes of death, expressed as a ratio of incidence rates in a high-dose group to the rate in a low-dose group. At doses some 200 times greater than experienced by the Hanford workers, the risk of leukemia was markedly elevated, and an increased risk is apparent for several other malignancies, none of which is elevated among the Hanford workers. On the basis of the criteria established from the study of the bomb survivors, the suggestion from the Hanford study that cancer is elevated due to radiation exposure is difficult to accept. Nevertheless, if one is willing to accept the worse case suggested by the various studies of the Hanford workers, namely that 5 percent of the cancer deaths were due to radiation exposure, what sort of risk does this imply? There are not enough data given in any of the reports to make an accurate calculation, but by limiting attention to the group that was hired in the first two years, approximately 6000 people, a crude estimate can be obtained. The incremental cancer rate, based on 5 percent of all cancers due to radiation exposure, appears to be of the order of eight additional deaths per year per 100,000 workers. This rate, combined with the appropriate accidental death rate, could be used as a true measure of risk for an industry or occupation.

This might constitute an excellent method of comparing industrial risks. For example, for a chemical industry any industrially related malignancies would be added to the deaths occurring on the job, and for the coal mining industry deaths due to black lung and emphysema would be added to the deaths due to accidents.

RADIUM TOXICITY

One method of evaluating the risk from plutonium is to examine human experience with a similar radionuclide. Radium, an alpha-emitting radionuclide like plutonium, deposits in bone where it remains, like plutonium, for a relatively long time. People have been exposed to radium since the turn of the century, providing experience we can draw on to establish criteria with which to evaluate the hazards posed by plutonium.

In a well-known example—the radium dial painting industry, which started around 1915 and was flourishing by 1920—radium was used as an activator for paints containing zinc sulfide. The women employed in painting the watch and clock dials were paid on a piecework basis. They found that they could paint accurately and rapidly if they used their lips to put a point on their paint brushes, and hence they inadvertently ingested radium as they painted. As a consequence of this ingestion, many of the young women soon showed toxic symptoms, which we now recognize to have been manifestations of acute radiation damage.

Following the pioneering work of the chief medical examiner of Essex County, New Jersey, Dr. Harrison Martland, the cause of disabling illnesses experienced by the dial painters was finally attributed to the radium in their paint. Once the cause was identified, in the year 1925, the tipping of the brushes in the mouth was discouraged, with the result that the quantity of radium deposited in the bodies of those who entered the industry after 1925 was much less than the quantity acquired by those who started in this industry before this date. Figure 8−2 is a graph plotting the quantity of radium estimated to have entered the body of each individual worker against the year of entry of that individual into the industry. Included are those who worked in the dial industry before 1950 whose radium body content has been measured and who acquired more than 0.1 μCi radium.

The figure demonstrates two relevant facts. First, the amount of radium acquired dropped markedly after the hazards of radium were understood. This drop is really quite remarkable, for there were no health physicists in the dial painting plants to enforce the rule against tipping the brush, there were no sophisticated instruments available to survey the work areas, and there were no government inspections to enforce the voluntary rules. The second fact is that the painting of dials with radium continued in spite of the recognized hazards of this material. True, there were not very many new employees during the period immediately after 1925, but by the time of World War II

Figure 8–2. Systemic Intake of Radium, in Units of μCi ^{226}Ra plus 2.5 times μCi ^{228}Ra, for Dial Workers who Entered This Industry before 1950, Plotted against Year of First Entry.

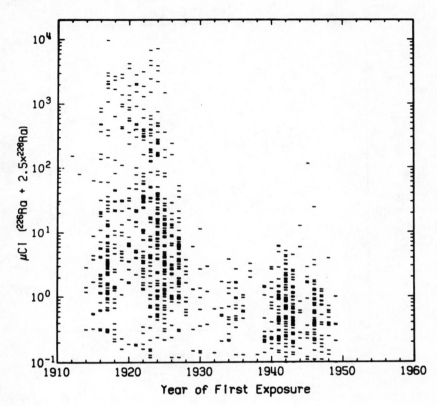

the industry experienced a renaissance because luminous instrument dials were needed for aircraft.

One of the best-known and most tragic consequences of the ingestion of radium is the induction of malignant tumors in bone. Radium behaves in the body in a manner analogous to calcium, and hence deposits primarily in bone. The alpha particles emitted by radium and its daughters bombard the cells adjacent to bone, a process which leads to the induction of malignant bone tumors. Figure 8–3 was constructed in the same manner as Figure 8–2, but shows the acquired radium and the starting date for only those radium dial painters who eventually developed a malignant cancer of the bone. Of all the dial workers who have been located, none who entered the

Figure 8–3. Systemic Intake of Radium for Dial Workers Who Ultimately Developed Bone Sarcomas Plotted against Year of Entry into this Industry.

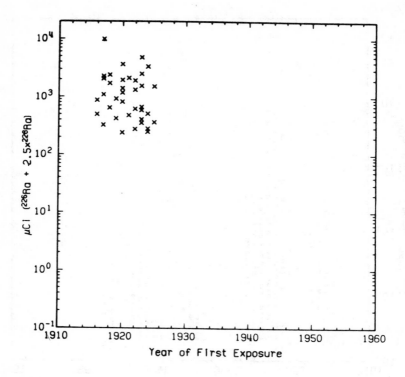

industry after 1925 have experienced a malignancy induced by radium.

From an analysis limited to those women who entered the radium dial painting industry before 1930, a quantitative description of bone sarcoma incidence as a function of radium intake has been obtained (Rowland, Stehney, and Lucas 1978). This population of 759 women, who started work at age 19.0 ± 4.5 years, experienced 38 bone sarcomas. Apparently all of these were induced by the radium they ingested, for not even one would have been expected from U.S. statistics, considering their age and survival. A linear relation cannot describe the observed values of initial intake and the bone sarcoma incidence, but these data are nicely fitted by a dose-squared-exponential function of the form

$$I = \beta D^2 e^{-\gamma D} \quad ,$$

where I is the number of bone sarcomas per person-year at risk and D is the quantity of radium that entered the blood. The values of β and γ were determined by a least-squares-fitting procedure, and were found to be

$$\beta = 6.8 \times 10^{-8} \ \frac{\text{bone sarcomas}}{\text{person-year} \ (\mu\text{Ci})^2}$$

and

$$\gamma = 1.1 \times 10^{-3} \ (\mu\text{Ci})^{-1} \ .$$

The fit of this dose-response expression to the actual data is shown in Figure 8−4. The decrease in bone cancer risk at the very highest doses does not imply that these levels are without risk; the radiation levels are so high that although few cancers have been seen, life expectancy is relatively short.

Figure 8−4. The Observed Number of Bone Sarcomas per Person-Year as a Function of Systemic Intake of Radium for Female Dial Workers First Employed Before 1930. The solid line is the best fit of an equation of the form I (bone sarcomas per person-year) = $\beta(\mu\text{Ci})^2 e^{-\gamma(\mu\text{Ci})}$, which in turn was the only form of the generalized dose-response function I = $(\alpha D + \beta D^2)e^{-\gamma D}$ that provided an acceptable fit to the data. From Rowland, Stehney, and Lucas (1978).

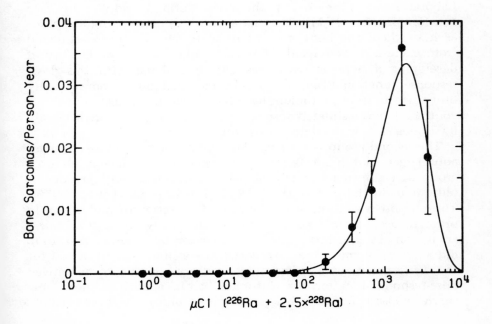

PLUTONIUM TOXICITY

In order to apply this dose-response relation derived from radium intake to plutonium, we need to find an equivalence between the two radionuclides. One such equivalence is developed in Appendix 8–1; it assumes that the relevant factor is the total energy delivered to cells on bone surfaces over the time from intake of the isotope to the mean time at which radium-induced bone tumors were diagnosed, less an assumed latent period. For each unit of radioactivity that enters the blood we find that plutonium delivers 570 times the energy to these cells as radium; the ratio is as high as this because plutonium is almost completely retained in the body, whereas radium is continually excreted.

The assumption is now made that the dose-response function for radium can be used for plutonium if the factor of 570 is introduced into the dose-response function, so it now appears as:

$$I = \beta (D \times 570)^2 \, e^{-\gamma \, \cdot \, D \, \cdot \, 570} \quad .$$

How can this function be tested? Two small groups of people who acquired plutonium in their bodies provide our only well-documented plutonium exposures. The first of these is a group of twenty-seven men who worked with plutonium at Los Alamos Scientific Laboratory when they were in the service during World War II (Voelz and Hempelmann 1977). They have been examined periodically; for each of them the body content of plutonium has been estimated from excreta measurements. For each individual the probability of developing a bone sarcoma has been calculated with this dose-response function (Table 8–9). The fact that no sarcomas or any other effects from plutonium have been seen in this group does not contradict the postulated dose-response function, for it predicts only 0.07 bone sarcoma to date in this group.

The second group is even smaller, but its members have considerably larger plutonium burdens. Information is available about a group of eighteen terminally ill patients who received injections of plutonium in the period 1945–1947 (Rowland and Durbin 1976). The purpose was to measure the rate of excretion of plutonium, in order to use such measurements to predict body contents of workers in contact with this element. Since seven members of this group did survive for more than five years, the assumed latent period for the development of a radiation-induced bone sarcoma, the derived dose-response function for plutonium can be applied to these long-surviving cases. The fact that no effects of plutonium have been ob-

Table 8-9. Bone Sarcomas Expected from Los Alamos Cases

Case	Body Burden (μCi)	Person-years at Risk	Bone Sarcomas Expected ($\times 10^{-3}$)
1	0.12	27.02	7.97
2	0.009	27.02	0.05
3	0.20	26.61	20.78
4	0.18	26.52	16.98
5	0.079	26.52	3.47
6	0.083	26.52	3.83
7	0.081	26.52	3.65
8	0.053	26.52	1.59
9	0.010	26.44	0.06
10	0.044	26.44	1.10
11	0.034	26.44	0.66
12	0.025	26.44	0.36
13	0.007	26.44	0.03
15	0.023	8.93	0.10
16	0.018	24.94	0.18
17	0.070	26.36	2.73
18	0.061	26.36	2.08
19	0.017	26.36	0.17
20	0.033	26.36	0.62
21	0.025	26.36	0.36
22	0.023	26.36	0.30
23	0.032	26.27	0.58
24	0.018	26.27	0.19
25	0.014	26.27	0.11
26	0.006	26.19	0.02
27	0.027	26.19	0.41
		Total	68.38

Body content and person-years at risk from data in Table 1, Voelz and Hempelmann (1977).

served to date is not contradicted by the prediction of 0.7 bone sarcoma in this group (Table 8-10).

It is apparent that the dose-response function for plutonium, derived from actual human experience with radium, is not contradicted by the limited human data available for plutonium exposures. True, it slightly overestimates the occurrence of bone sarcomas in the only human experience available, but the overestimation, 0.8 bone tumor expected in the two groups, none observed, is certainly not statistically significant.

What predictions can be made from this plutonium dose-response function? At the generally accepted maximum permissible body content for workers in the nuclear industry, 40 nCi, the incremental risk is 4.2 bone sarcomas per year per 100,000 workers (see Appendix

Table 8–10.   Bone Sarcomas Expected from Plutonium Injection Cases

Case	Plutonium Injection (μCi)	Person-years at Risk	Bone Sarcomas Expected
Cal–1	3.5	15.67	0.52
HP–1	0.28	9.25	0.01
HP–3	0.30	27.70	0.05
HP–6	0.33	27.52	0.05
HP–8	0.40	24.73	0.07
HP–10	0.38	5.89	0.02
HP–12	0.29	3.01	0.01
		Total	0.72

Extracted from Table 1, Rowland and Durbin (1976).

8–2). This, of course, is under the extreme assumption that every worker acquires the maximum permissible level of plutonium.

As an alternative method of examining the hazards that the use of plutonium engenders, consider the following hypothetical scenario. Assume that two alpha-emitting radioelements were available to activate the paint, plutonium as well as radium, when the dial painting industry started. Further, assume that the decision was made to use $^{239}PuO_2$ as the activator, instead of radium. What would have been the consequences of this decision for the hypothetical plutonium dial painting industry?

In Appendix 8–3, it is shown that in place of each unit of radium, 3.4 units of plutonium activity would have been required to activate the paint. The difference arises because radium has daughter products that contribute more alpha particles than radium itself. Plutonium, in contrast, has no effective daughters; therefore, more plutonium must be used to supply the necessary alpha particles.

Taking into account the larger quantity of plutonium that would have been ingested by the dial painters, the much poorer absorption of plutonium through the intestine, and the much greater toxicity per unit of plutonium activity, the risk for each of the 759 female dial workers, now assumed to have ingested plutonium instead of radium, has been calculated. The prediction for the dose-response function for bone sarcomas from plutonium is that six such malignancies would have been observed in this population, in contrast to the thirty-eight bone sarcomas that were actually induced by radium.

At this point one might suggest that the risk from plutonium at low doses is not from bone sarcomas, for their incidence varies with the square of the intake, but from lung cancer. Is there any relevant human experience that can be used to evaluate this risk?

At the present time only negative information exists. For example, it is known that the plutonium workers at Los Alamos inhaled plutonium, for detectable levels of plutonium were often found when nasal swipes were made at the end of their working hours. Their estimated body burdens, however, have been obtained from measurements of plutonium in urine and do not include an estimate of the plutonium in the lung. Plutonium deposited in the lung enters the blood very slowly, particularly if it is in an insoluble form. The actual quantity of plutonium in the lungs of these men is unknown.

Therefore these former workers should be considered to be at risk of developing lung malignancies, but none has been observed. From this negative information we can only conclude that this risk is low enough so that no lung cancers have been induced in these early plutonium workers.

It is hoped that an ongoing study of another group of workers may lead to an understanding of the relation between the quantity of an alpha-emitting isotope in the lung and the subsequent induction of lung malignancies. These are workers who were exposed to dust containing thorium in a plant that extracted rare earths and thorium from monazite sands over a period of forty years. It is too early to have the necessary detailed understanding of the lung doses experienced by these workers from their exposure to thorium, for the study has just gotten underway. Nevertheless, since the number of employees is quite large, we have already developed from a study of death certificates a preliminary description of the causes of death in this group.

To date, nearly fifty former workers have been examined at Argonne, and all but two have been found to contain measurable levels of thorium. The levels are small, of the order of a few nanocuries, but since thorium-232 is the parent of five alpha-emitting daughters, a long-term deposit of thorium could produce a total of six alpha particles per thorium atom decay. Thus a lung burden of 3 nCi of thorium-232 might produce a total alpha-emission equivalent of 18 nCi of plutonium. Since the maximum permissible lung burden of plutonium-239 is 16 nCi, we conclude that these thorium workers have experienced lung doses comparable to the maximum allowed from plutonium.

Table 8−11 shows a preliminary listing of some of their causes of death, determined from an analysis of 529 certificated deaths from a cohort of 2986 male employees of this plant. The expected number of deaths has been obtained from age- and time-specific rates for U.S. white males. It is apparent that the total number of deaths is greater than expected and that the total number of malignancies is

Table 8–11. Observed and Expected Deaths from Specific Causes for 2986 Male Employees of an Extraction Plant

ICD No.	(7th Revision) and Cause	Deaths		Obs./Exp.
		Observed	Expected	
	All causes	529	485.8	1.09
140–205	All malignant neoplasms	103	81.1	1.27[a]
162	Lung cancer	31	20.84	1.49[a]
810	Motor vehicle accidents	41	22.9	1.79[b]

[a]$P < 0.05$ (chi-square, Mantel–Haenzel).
[b]$P < 0.001$.

also greater than expected. The dusty working conditions in this extraction plant would be expected to lead to inhalation and perhaps ingestion of the dust. Inhalation would expose the workers to alpha radiation of the lung from the long-lived thorium as well as the hazards from the other nonradioactive components of the dust. Since thorium is very poorly absorbed through the intestine, very little is expected to enter the systemic circulation following ingestion.

Table 8–11 indicates that there is a significant increase in lung cancers among these workers. Although the data in this table indicate that these workers had higher than expected death, malignancy, and lung cancer incidence, a note of warning is interjected by the last entry in the table. The most significant departure from expected death rates was due to motor vehicle accidents, which may be an indication that this group of employees differs from the norm for U.S. white males.

It is generally recognized that an induction period is required for radiation-induced malignancies. A ten-year latent period is usually used for solid tumors (with the exception of bone sarcomas, for which five years is the normal latent period). Therefore, if some of the lung cancers experienced by these workers were induced by the alpha particles from thorium deposited in the lung, these should appear after a ten-year latent period. Furthermore, it would appear logical to assume that the longer an employee worked in this dusty plant, the more thorium he might inhale, and the higher his risk of experiencing a radiation-induced lung cancer.

These two concepts are examined in Table 8–12, in which lung cancers that resulted in death ten or more years after first employment are tabulated for four length-of-employment groups. First of all, the ratio of observed to expected lung cancers for the entire group is seen to have changed very little by removing the first ten

Table 8–12. Observed and Expected Deaths from Lung Cancer Occurring after Ten or More Years of Employment in an Extraction Plant

Duration of Employment	Workers in Category (Male Employees)	Lung Cancers after Ten Years (Male Employees)		
		Observed	Expected	Obs./Exp.
Less than 2 months	1222	10	6.13	1.63
2–11 months	1094	5	5.46	0.92
12–35 months	352	5	1.76	2.84[a]
36 or more months	318	3	2.89	1.04
All employees	2986	23	16.24	1.42

[a]$P < 0.05$ (chi-square, Mantel–Haenzel).

years of experience, actually decreasing from 1.49 to 1.42, instead of increasing as might be expected. Second, the very shortest exposure group, those who worked less than two months, are seen to have a high lung cancer rate, as did those who worked one to three years. Those who worked from two to twelve months, or more than three years, have, in contrast, a very normal lung cancer rate.

Of the approximately seven excess lung cancer deaths observed in the workers at this extraction plant, four appeared in those who were employed only a very short time. The remainder, instead of being distributed according to length of exposure to dust, all appeared in the group that worked from one to three years. If we assume, nevertheless, that these three were the result of exposure to dust at the plant, we can calculate a crude risk. There were 26,522 person-years of experience among the plant employees, when the summation of years at risk started ten years after their employment date. The three lung cancer deaths per 26,552 person-years is a rate of 11.3 additional deaths by lung cancer per year per 100,000 people. This is a relatively high risk, almost as large as the average overall accident rate of fifteen deaths per year per 100,000 people suggested as the division between safe and risky occupations. If the risk is due entirely to the plant (thorium or other constituents of the dust), it suggests that a measurable hazard does exist from the exposures that occurred. The evidence, however, does not necessarily point toward thorium as the cause of the elevation in lung cancer deaths.

Despite the fact that its members were exposed to a mixture of materials, this group does appear to have potential for the evaluation of the hazards of alpha-particle irradiation of the human lung. Measurements of the actual lung doses and evaluation of some confounding variables, such as smoking histories and exposures to other carcinogenic materials (at this plant or others), would permit calculation of risk as a function of dose. Until this information is available, we can make only the following statement. There may be an elevation of lung cancer deaths among workers in a plant where they were exposed to dusts containing thorium. The rate may be as large as eleven lung cancer deaths per year per 100,000 exposed. The lung doses are unknown; the suggestion from the few cases on whom measurements have been made indicate that measurable levels of thorium are present in most cases.

CONCLUSIONS

There is no simple method of expressing a concept such as "toxicity." The toxicity of a material depends upon its form (is it a solid, liquid, or gas?); the method of entry into the body (does it enter by ingestion, inhalation, or abrasion?); and the individual it enters (is he or she young, old, sick, well, a heavy cigarette smoker, or a consumer of alcohol?). Thus, statements such as "Plutonium is the most toxic material known to man" are not only incorrect; they are meaningless.

By taking the one aspect of plutonium I believe can be quantified in a meaningful manner, the induction of malignant bone sarcomas in human beings, the incremental risk presented to those who have been exposed to this radioelement has been shown to be quite small. By means of the hypothetical exposure of radium dial workers to plutonium instead of radium, I suggest that an industrywide exposure to plutonium would have been less harmful than the actual exposure to radium, in spite of the fact that plutonium appears to be far more toxic than radium. Furthermore, the observation that, after radium was recognized as a toxic material, it no longer was a hazard in the dial industry, clearly demonstrates that hazardous radioactive materials can be handled safely. This has been demonstrated again with plutonium, for even though much larger quantities have been handled by many more individuals, we have no demonstrable cases of malignancies induced by this hazardous material.

The absence of clearly demonstrable excesses of lung cancer in the large group of long-term thorium workers is reassuring, for it may imply that low doses to the lung have a very low probability of inducing lung cancers. I conclude that plutonium, a highly toxic material, can be handled safely, as are many other important but hazardous materials in our daily lives.

APPENDIX 8-1

In a study of radium-induced malignancies in a population of 759 female dial workers exposed before 1930, the mean appearance time (± SD) of the 38 bone sarcomas in this population was 27.3 ± 4.5 years (Rowland, Stehney, and Lucas 1978). If a latent period of five years is assumed, the cumulative dose to the endosteal cells of bone for the first twenty-two years after acquisition is the relevant mean measure of dose. The value of this dose for unit quantities of plutonium-239 and radium-226 will then provide a measure of the ratio of toxicity (for the induction of bone sarcomas) for these two radioisotopes.

The calculation for plutonium-239 assumes that, of the plutonium that reaches the systemic circulation, 45 percent deposits and remains on bone surfaces, from which it is released with a mean life of 144 years. The cumulative average bone dose is given by the equation:

$$D(T) = f \cdot I \cdot Q \ \int_0^T e^{-\lambda t} \cdot dt$$

where f is the fraction of isotope in the organ, 0.45 in this case (ICRP 1972); I is the amount of the isotope entering the systemic circulation (for convenience we use 1 μCi); Q is the dose rate per unit of activity, 0.0776 rad/μCi-day for 3400 g female bone tissue; T is the time, in days, over which the dose is calculated (22 years × 365 days); and λ is the effective elimination constant (1.91 × 10^{-5} day^{-1}) (ICRP 1972). So, D(22 years) = 260 rads.

The endosteal dose from surface-deposited plutonium-239 has been shown to be 12.8 times the average bone dose (Marshall, Groer, and Schlenker 1978), or 3330 rads over a period of twenty-two years from 1 μCi of plutonium-239 entering the systemic circulation.

The calculation for radium-226 utilizes the model of ICRP 20 (ICRP 1973), where the average bone dose is given by

$$D(T) = I \cdot Q \ \int_0^T Rdt$$

where I is assumed to be 1 μCi; Q is 0.185 rad/μCi-day for 35 percent radon retention and 3400 g female bone tissue; T = 22 years × 365 days, and $\int_0^{22 \text{ yr}} Rdt$ = 70 days (ICRP, 1973). So, D(22 years) = 13 rads.

The endosteal dose from volume-deposited radium–226 has been shown to be 0.45 times the average bone dose (Marshall, Groer, and Schlenker 1978), or 5.85 rads over a period of twenty-two years from 1 μCi radium–226 entering the systemic circulation.

Therefore, per μCi entering the systemic circulation, plutonium–239 delivers a dose 3330/5.85, or 570 times as large as the dose from radium–226.

The dose-response function for bone sarcoma induction in the group of 759 female dial workers is

$$I = \beta D^2 e^{-\gamma D}$$

where I is bone sarcomas per person-year; D is the systemic intake in μCi; β is 6.8×10^{-8} bone sarcomas (person-year)$^{-1}$ (μCi)$^{-2}$; and $\gamma = 1.1 \times 10^{-3}$ (μCi)$^{-1}$.

For plutonium, this is modified by increasing the systemic intake values by the toxicity ratio, 570. That is, for bone sarcoma induced by plutonium–239,

$$I = \beta(570D)^2 e^{-\gamma(570D)}$$

or

$$I = 2.2 \times 10^{-2} D^2 e^{-0.6D}.$$

APPENDIX 8–2

Risk at the Maximum Permissible Body Burden

According to ICRP 19 (ICRP 1972), 90 percent of the plutonium that enters the blood is long retained in the body, divided between bone and liver. Therefore, an intake of 44.5 nCi plutonium–239 would result in an initial 40 nCi body burden.

Using the dose-response function derived in Appendix 8–1,

$$I = 2.2 \times 10^{-2} D^2 e^{-0.6D} ;$$

with D = 0.0445 μCi ,

$$I = 4.24 \times 10^{-5} \frac{\text{bone sarcoma}}{\text{person-year}}.$$

This function predicts, after a five-year latent period, 4.2 bone sarcomas per year per 100,000 people who acquired the maximum permissible body burden of plutonium–239.

APPENDIX 8–3

The paint used in the dial industry contained, on the average, about 30 μCi radium per gram of paint (range 3–300 μCi/g). For paint containing only radium–226 the alpha activity in each gram of paint was 2.26×10^8 α/min, on the assumption that 20 percent of the radon was lost from the paint. For paint containing equal activities of radium–226 and radium–228, with the radium–228 aged for about two and one-half years to allow growth of the alpha-emitting daughters, the total alpha activity is also 2.26×10^8 α/min, assuming no emanation of thoron.

If plutonium is to be substituted for radium, the required 2.26×10^8 α/min implies 102 μCi plutonium–239 per gram of paint, or 3.4 μCi in place of each μCi of radium.

Since the half-life of plutonium–239 is 24,000 years, implying a relatively high mass per unit of activity, one might wonder if the required mass of ^{239}PuO$_2$ could be added to the paint without changing its characteristics. The specific activity of plutonium–239 is 6.1×10^{-2} Ci/g, so 102 μCi would weigh only 1.7 mg; the required quantity of PuO$_2$ would contribute less than 2 mg per each gram of paint.

The systemic intake for each of 759 measured female dial workers who started before 1930 has been estimated. Since 20 percent of ingested radium is absorbed (ICRP 1973), the total quantity of radium ingested by each was five times the systemic intake. The hypothetical plutonium ingestion for each case is 3.4 times the radium ingestion.

The absorption of insoluble plutonium is generally considered to be only 3×10^{-5} of the quantity ingested; use of this factor then provides the hypothetical systemic intake of plutonium for each case. The dose-response function derived in Appendix 8–1 was then applied to each of the 759 female dial workers; the sum over all cases was 6.1 bone sarcomas expected from the individual plutonium systemic intake values.

REFERENCES

International Commission on Radiological Protection (ICRP). 1972. *The Metabolism of Compounds of Plutonium and Other Actinides.* ICRP Publication 19. Oxford, England: Pergamon Press.

_____. 1973. *Alkaline Earth Metabolism in Adult Man.* ICRP Publication 20. Oxford, England: Pergamon Press.

Kneale, George W.; Alice Stewart; and Thomas F. Mancuso. 1978. "Reanalysis of Data Relating to the Hanford Study of the Cancer Risks of Radiation Workers." In *Proceedings of the International Symposium on the Late Biological*

Effects of Ionizing Radiation (Vienna, Austria, March 13–17, 1978). Vienna: International Atomic Energy Agency.

Mancuso, Thomas F.; Alice Stewart; and George Kneale. 1977. "Radiation Exposures of Hanford Workers Dying from Cancer and Other Causes." *Health Phys.* 33: 369–85.

Marks, Sidney; Ethel S. Gilbert; and Bryce D. Breitenstein. 1978. "Cancer Mortality in Hanford Workers." In *Proceedings of the International Symposium on the Late Biological Effects of Ionizing Radiation* (Vienna, Austria, March 13–17, 1978). Vienna: International Atomic Energy Agency.

Marshall, J.H.; P.G. Groer; and R.A. Schlenker. 1978. "The Dose to Endosteal Cells from Radium–224," *Health Phys.* 35 (July): 91–102.

Moriyama, Iwao M. 1978. *Capsule Summary of Results of Radiation Studies on Hiroshima and Nagasaki Atomic Bomb Survivors, 1945–75.* Technical Report RERF TR 5–77. Japan: Radiation Effects Research Foundation.

National Safety Council. 1976. *Accident Facts.*

Rowland, R.E., and P.W. Durbin. 1976. "Survival, Causes of Death, and Estimated Tissue Doses in a Group of Human Beings Injected with Plutonium." In *The Health Effects of Plutonium and Radium*, proceedings of a symposium, Sun Valley, Idaho, October 6–9, 1975, W.S.S. Jee, ed. Salt Lake City: The J.W. Press.

Rowland, R.E.; A.F. Stehney; and H.F. Lucas, Jr. 1978. "Dose-Response Relationships for Female Radium Dial Workers." *Radiat. Res.* 76: 368–83.

Sanders, B.S. 1978. "Low-Level Radiation and Cancer Deaths." *Health Phys.* 34: 521–538.

U.S. Bureau of the Census. 1973. *Vital Statistics of the United States.* Vol II: *Mortality.* Pt. A.

Voelz, G.L., and L.H. Hempelmann. 1977. "Health of Los Alamos Plutonium Workers." In *Biomedical and Environmental Research Program of the LASL Health Division, January–December 1976.* Report no. LA–6898–PR, D.F. Petersen and E.M. Sullivan, compilers. Los Alamos, N.M.: Los Alamos Scientific Laboratory.

※　*Chapter 9*

Discussion of Plutonium Toxicity

Charles W. Mays

In discussing Dr. Rowland's paper, "Some Comments on the Toxicity of Plutonium" (Chapter 8), let me first state our points of agreement:

1. Plutonium is *not* the most toxic material known to mankind. It is toxic because of its radioactivity. The long half-life of 24,000 years for plutonium–239 makes it much less radioactive and, therefore, much less toxic per microgram than short-lived alpha-emitters, such as thorium–228 which has a half-life of 1.9 years. Fortunately, plutonium is poorly absorbed through both the digestive tract and the skin, leaving inhalation as the most important route of intake.
2. Detailed study of the radium dial painters is extremely valuable in that it provides direct information on the effects of α-particle irradiation in the human skeleton. It is my personal belief that the follow-up of these persons should continue until nearly all of them have died.
3. I have no quarrel with Dr. Rowland's dose-squared fit to the existing bone sarcoma data in the radium dial painters of known dose.

Now, I want to explore an alternative way of looking at the problem. Can we be certain that the dose-squared response for the special case of the radium dial painters also applies to plutonium?

I made Figure 9–1 many years ago to show my students three possible dose-response curves to these illustrative data. One curve shows

Figure 9−1. Dose-Response Examples

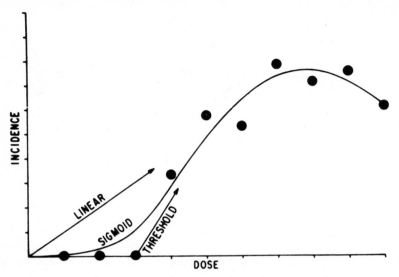

the "linear" possibility, in which incidence of radiation-induced disease increases in direct proportion to the radiation dose. Another is the "threshold" possibility, in which nothing happens until a threshold dose is exceeded. Then, what used to be my pet is the "sigmoid" response, curving upwards, perhaps initially as a dose-square (or some other mathematical function), reaching a peak, and then finally bending down, because of the overkill of cells or, perhaps, the overkill of people. The sigmoid dose-response describes very well the incidence of bone sarcomas in the radium dial painters exposed to α-particle radiation. In addition, it is the *typical* dose-response in mammals exposed to *sparsely* ionizing radiations from X rays, β-particles, and γ-rays.

Figure 9−2 shows that with recent evidence for *densely* ionizing radiation resulting from α-particles and fast neutrons, however, there is another possibility in addition to the linear and concave-upwards responses. This is the possibility of a concave-downwards response, in which the risk per rad at very low doses might be even higher than derived from the linear model.

Before concentrating on bone sarcomas, I will mention that for other human malignancies induced by densely ionizing radiation, the dose-response for neutron-induced leukemias at Hiroshima appears to be more or less linear (Rossi and Mays 1978), the dose-response for lung cancers induced by inhaled alpha-emitters in uranium miners

Figure 9–2. Dose-Response Possibilities *(At low doses of densely ionized radiation)*

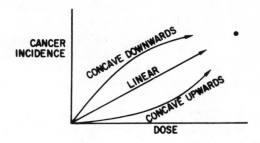

can be regarded either as approximately linear or perhaps concave downwards (Svec, Kunz, and Placek 1976), and in the radium dial painters themselves the dose-response for the head sinus carcinomas is best described by a linear relation (Rowland, Stehney, and Lucas 1978), although the response for their bone sarcomas appears to be concave upwards (Rowland, Stehney, and Lucas 1978).

I am going to show you some examples of experimental data on bone sarcoma induction by α-emitters. A few are concave upwards; one is concave downwards; but in most the possibility of an approximately linear relation cannot be excluded.

I present the results in terms of the "incidence" (that is, proportion of the individuals who develop bone sarcomas). For the animal studies shown on the slides, all of the animals have died, and, hence, the incidence is their final incidence. I have used incidence because it is a simple concept that is easily explained and widely understood. It enables information to be derived from a number of important studies in which the individual survival times are not available for each subject. Such information is needed for more sophisticated analysis, such as by Kaplan-Meier statistics (Kaplan and Meier 1958), which is intended to correct for competing risks. This type of analysis is valuable and should be done when possible, but I must point out that in many instances, the information is not readily available to make this type of analysis. In many studies with bone-seeking radionuclides, bone sarcomas (defined as osteosarcomas, chondrosarcomas, and fibrosarcomas of bone) are the main radiation-induced cause of death, and thus, for healthy animals, in which the increasing risk per year from "natural" causes does not fluctuate widely from group to group, the unreliability caused by competing risks is not excessive. (Please see "Response by Charles W. Mays to comments by Alvin M. Weinberg on Shape of the Dose-Response Curve" in the post-symposium dialogue in the Appendix of this volume for more details.)

For long-lived bone-seeking radionuclides, the appearance times for bone sarcomas tend to decrease as the dose increases. Thus, a bone sarcoma at a high dose level tends to subtract more lifespan than a bone sarcoma at a low dose level. But at the low dose levels of greatest importance for radiation protection, the average lifespans of the low dose groups tend to be rather similar due to the low frequency of radiation-induced bone sarcomas, and in laboratory animals there is a tendency for these tumors to appear late in the lifespan.

In general, it is difficult to describe the entire dose-response relation by simple equations. This may be because different mechanisms and promoting factors could operate at high doses, but be relatively insignificant at low doses. For example, while simple relations, such as linear or dose square, may well describe the response at low doses, they can, if unmodified, predict impossible answers at high doses, such as radiation-induced diseases occurring in more than 100 percent of the individuals at risk! I have mainly concentrated on the response at low doses. My purpose is to show that a variety of dose-response relations have been observed for bone sarcoma induction by low doses of α-radiation, and that while Bob Rowland's dose-squared relation makes an excellent fit to the existing data on the radium dial painters, it seems unlikely to give a best fit for many of the other examples, and hence it may not best represent the general dose-response for α-radiation.

DOSE-RESPONSE RELATION FOR BONE SARCOMA INDUCTION BY α-EMITTERS

Figure 9−3 shows bone sarcoma induction by radium−226 (half life = 1600 years) in the CFI mice of Miriam Finkel (Finkel, Biskis, and Jinkins 1969). Average skeletal dose was calculated up to 100 days before tumor recognition; the final 100 days of radiation was regarded "wasted" with respect to the induction of a visible tumor (Mays and Lloyd 1972). The incidence of radiation-induced bone sarcomas appears to increase linearly with dose up to about 1000 rad. In the 1436 mice below 300 rad, 115 cases of bone sarcoma were observed, in excellent agreement with the 109 cases predicted (92 radiation-induced plus 17 naturally expected). Please note that radium−226 was also involved in the irradiation of the contaminated dial painters, many of whom were also exposed to radium−228.

Figure 9−4 shows the results in my C57BL mice of the Tom Dougherty strain (Mays et al. 1976). For radium−226, the response can be regarded as either linear or slightly concave upwards. The risk

Figure 9−3. ^{226}Ra-Induced Bone Sarcomas in Female Mice
(*Finkel, Biskis, and Jinkins 1969*)

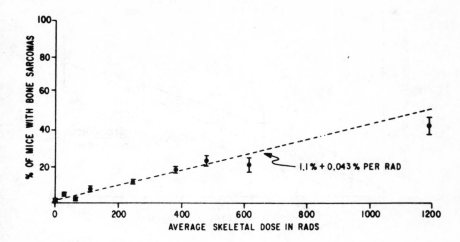

Figure 9−4. Bone Sarcoma Incidence in C57 BL/Do Mice *(Both Sexes)*

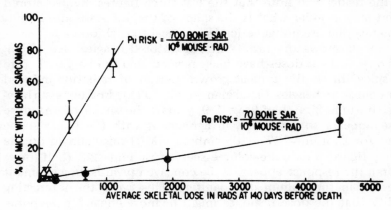

from plutonium–239 (half life = 24,000 years) appears to rise more
or less linearly with dose.

Figure 9−5 gives results from the NMRI mice of Luz et al. Thorium–227 is a short-lived α-emitter (half-life = 19 days) that, like
plutonium, deposits on bone surfaces. The incidence seems to increase linearly with dose. In the case of radium–224, in which the
short half-life of 3.6 days causes an important part of the α-particles
to be emitted from bone surfaces, we have an example of a concave-
downwards response.

Figure 9—5. Osteosarcomas in Female NMRI Mice *(Luz et al.1976: 171)*
(Biological and Environmental Effects of Low-Level Radiation)

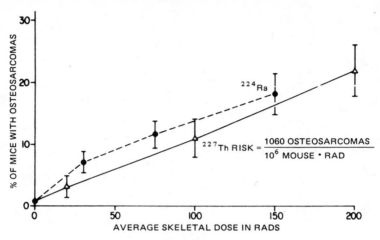

But the reader will note that the last three figures involve inbred mice and may wonder what is the shape of the dose-response in outbred animals that are not genetically identical to each other.

Figure 9—6 shows the results in our outbred beagles (Jee 1978). The average skeletal doses have been recalculated to one year before death, since this is the typical growth period of radiation-induced osteosarcomas in beagles (Thurman et al. 1971). For bone-surface-seeking thorium–228 (half-life = 1.9 years), the shape of the dose response appears to be linear, in agreement with the linear shape observed for thorium–227 in the inbred NMRI mice shown on the preceding figure. For bone-volume-seeking radium–228 (half-life = 5.8 years), the response appears to be concave upwards, in agreement with results in the radium dial painters. This raises the interesting possibility that the dose response might be more linear for the bone-surface-seekers (such as plutonium) than for the bone-volume-seekers (such as radium–226 and radium–228), in agreement with theoretical predictions by John Marshall and Peter Groer (1977). On the assumption that hits on two targets within a cell nucleus are necessary to induce bone cancer, Marshall and Groer point out that the flattened endosteal cell is more likely to be hit in two targets by a grazing α-particle emitted from the bone surface than by an approximately perpendicular α-particle emitted from within bone volume.

An alternative hypothesis exists that the damage from an α-particle is so severe that only a single target needs be hit to induce cancer,

Figure 9−6. Bone Sarcoma Induction by ^{228}Th and ^{228}Ra in Beagles *(March 1978)*

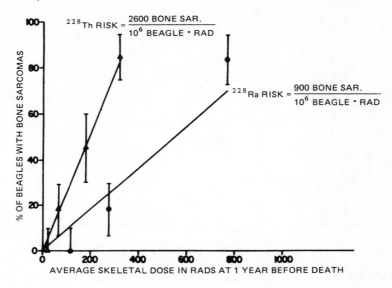

in which case the expected response at low doses for all α-emitters could be linear.

Figure 9−7 shows results as of 1975 for our beagles injected with plutonium-239 or radium-226 (Mays et al. 1976). As of 1975, these results supported the concept of an approximately linear response for bone-surface-seeking plutonium-239, but a concave-upwards response for bone-volume-seeking radium-226, the latter agreeing with results for radium-228 in beagles (see Figure 9−6), and for dial painters irradiated with radium-226 and radium-228. But in laboratory animals we have an option not ethically permitted in irradiated human populations: we can repeat the experiment. Indeed, we did this during 1964−1974 in young adult beagles at our standard injection age of about seventeen months, starting at the lowest doses in our original (now completed) studies of plutonium-239 and radium-226 in beagles, and extending these newer studies to even lower levels. These newer studies are still in progress and will require about ten more years until the last dog has died. But already we have seen bone sarcomas below the lowest levels at which these tumors had occurred in our original study (see data of Tables A1 and A2 in "Response by Charles W. Mays to Comments by Alvin Weinberg" in the Appendix). At the present time, we cannot rule out the distinct possibility that the dose response for bone sarcomas is approxi-

Figure 9—7. Bone Sarcoma Induction in Beagles *(October 1975)*

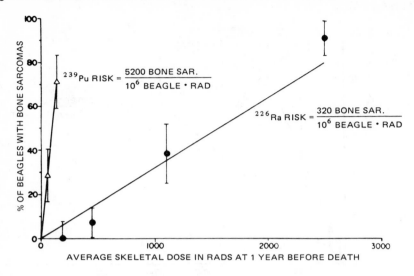

AVERAGE SKELETAL DOSE IN RADS AT 1 YEAR BEFORE DEATH

mately linear in beagles at low doses, both for plutonium–239 and radium–226, although the final results might best support a response that is concave downwards, or conversely, concave upwards.

We probably can never know whether the experience of the radium dial painters, if repeated, would replicate a dose-squared response. But there exists a different set of human data on about 900 German patients given repeated injections of radium–224, who are being followed by Professor Heinz Spiess (see Figure 9—8). The short, 3.6-day half-life of radium–224 minimizes its diffusion into bone volume and causes an important fraction of the α-particles to be emitted from bone surfaces. This gives a local distribution of dose somewhat similar to that from plutonium. Already, fifty-four cases of bone sarcoma have occurred among these patients, a value similar to that of fifty-six cases of bone sarcoma among the U.S. subjects of known dose from long-lived radium. In the radium–224 patients, the responses for both the juveniles and adults seemed to be approximately linear (Spiess and Mays 1971), at least up to 1000 rad, although we now know that the fit is better for a linear coefficient multiplied by an injection span factor, which allows for increasing effectiveness per rad with greater protraction of the span of α-particle irradiation (Spiess and Mays 1973; Mays, Spiess, and Gerspach 1978). The enhanced effectiveness with greater protraction of α-particle irradiation has been confirmed in the well-controlled

Figure 9–8. ^{224}Ra-Induced Bone Sarcomas in Humans *(Spiess and Mays 1971)*

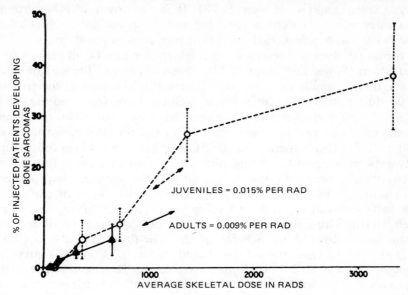

mouse experiments of Müller et al (1978), which show that the protraction effect is real and not a "dose-squared" effect in disguise.

Most of the radium–224 patients in the Spiess follow-up received
high doses (average skeletal doses exceeding 90 rad, the lowest dose
at which a bone sarcoma has occurred in the Spiess study). Therefore, it is appropriate to question how well our protraction-modified
linear relation, which was mainly based on high-dose data, applies
at lower doses. Before his untimely death, Fritz Schales was able
to obtain additional follow-up information on about 1170 adult
radium–224 patients with average skeletal doses below 90 rad
(Schales 1978; Mays 1978). At last contact, these patients had been
followed about 12 years after radium–224 injection, and the mean of
their average skeletal doses was about 56 rad. Among these low-dose
patients, two cases of "skeletal" sarcoma are known to have appeared, both at average skeletal doses of about 67 rad. One was a
fibrosarcoma of the iliosacral region (Wick 1978), while the other
was diagnosed as a reticulum cell sarcoma of bone marrow (Mays
1978). The fibrosarcoma was probably radiation-induced since this
type of bone tumor has occurred both in the radium–224 patients of
the Spiess series and in the American radium dial painters. A radiation induction of the other sarcoma is less certain since only one case

of reticulum cell sarcoma has been diagnosed in the Spiess series
(Mays, Spiess, and Gerspach 1978). It is unknown whether other
bone sarcomas may exist among the low-dose patients, which, for
reasons such as medical-legal complications, remain unreported.

From our linear dose-response relation for adults, modified for
protraction (Spiess and Mays 1973), about 3.5 cases of bone sarcoma
are *eventually* predicted to appear during the lifespans of the traced
1170 adult patients whose average skeletal doses from radium–224
are below 90 rad. But these patients have only been followed an
average of about twelve years, whereas in the Spiess series, the total
adult patients (both living and dead) now have an average follow-up
of twenty years, and the living adults now have an average follow-up
of twenty-six years. In the Spiess adults, the sarcoma appearance
times range from five to twenty-two years, with eleven of the eighteen total cases appearing by twelve years (Mays, Spiess, and Gerspach 1978). Thus, the total number of bone sarcomas predicted, as
of the last follow-up by Schales of the low-dose patients, is (3.5)
(11/18) = 2.14 cases radiation-induced, plus 0.14 cases naturally
expected, or about 2.3 total predicted cases. This agrees well with
the two cases presently observed. However, with scientific caution, I
remind the reader that this does not prove the exact correctness of
our linear model as modified to include the protraction effect, but
only that it is not in disagreement with present evidence.

Figure 9−9 was not shown at the Argonne symposium, but contains highly relevant information on the dose-response to low doses

Figure 9−9. Plutonium-Induced Bone Sarcomas in Rats *(Buldakov and
Lyubchanskiy 1970)*

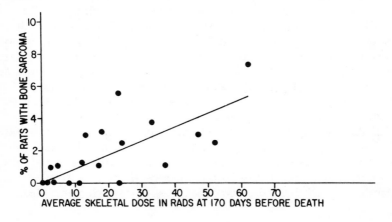

of plutonium-239 in the 2208 Wistar-derived rats of Buldakov and Lyubchanskiy (1970). I have recalculated the doses to 170 days before death to exclude radiation received too late to induce an observable osteosarcoma. In the 1457 rats with average skeletal doses between 0.03 and 18 rad, eleven cases of bone sarcoma were observed, whereas the indicated linear relation gives a predicted expectation of eight cases. No bone tumors were observed in 1323 control rats, signifying a very low natural incidence.

Table 9–1 is a summary of the indicated shapes of dose-response in eleven examples of bone sarcoma induction by α-particles, as of year 1976. In seven examples, the response then appeared approximately linear, in one example, concave downwards, and in three examples, concave upwards. Our recent data suggest, however, that the final response for radium-226 in beagles might change from concave upwards to either approximately linear, or even concave downwards.

I agree that the existing data for bone sarcoma induction by radium-226 and radium-228 in the dial painters is well represented by a concave-upwards response. But if this is taken as the universal response, then *why are there so many exceptions?* I feel that the weight of present evidence suggests that it is more *realistic* to assume that the response at low doses of plutonium is *approximately linear*. However, I am quite prepared to change my opinion if it is no longer supported by the weight of future evidence.

Table 9–1. Bone Sarcoma Induction versus α–Dose As of 1976

Approximately Linear

1. ^{224}Ra in Humans 0–1000 Rad (Spiess & Mays 1973)
2. ^{239}Pu in Beagles 0–140 Rad (Mays et al 1976)
3. ^{228}Th in Beagles 0–320 Rad (Mays, unpublished)
4. ^{239}Pu in Rats 0–60 Rad (Buldakov et al 1970)
5. ^{226}Ra in Mice 0–500 Rad (M. Finkel et al 1969)
6. ^{227}Th in Mice 0–200 Rad (Luz et al 1976)
7. ^{239}Pu in Mice 0–1000 Rad (Mays et al 1976)

Concave Upwards

1. ^{226}Ra & ^{228}Ra in Man 0–2000 Rad (Rowland & Stehney 1974)
2. ^{226}Ra in Beagles 0–2500 Rad (Mays et al 1976)
3. ^{228}Ra in Beagles 0–800 Rad (Mays, unpublished)

Concave Downwards

1. ^{224}Ra in Mice 0–150 Rad (Luz et al 1976)

ESTIMATED RISK TO HUMAN BEINGS
FROM PLUTONIUM

Table 9−2 shows the main types of cancer likely to be induced by plutonium. The usual intake of plutonium is by inhalation, and irradiation of the lung can produce lung carcinomas. From the lung, the plutonium slowly translocates to the liver, where it can produce liver cancers, and to the bone, where it can produce bone sarcomas. Fortunately, because of the tragic experience of the radium dial painters, the safety precautions taken in the plutonium industry have prevented a massive epidemic of plutonium-induced cancers. Dr. Rowland is correct: at present, there is no direct evidence that any individual case of cancer in humans can be positively proven to have been induced by plutonium. But I believe that this is due to the care taken to limit the intake of plutonium rather than some presumption that plutonium within the body might be nontoxic. Human experience shows that α-particle irradiation can cause cancer in man, and animal results show that plutonium–239 is more toxic than radium–226, in terms of average skeletal dose (Mays et al. 1976). With the linear hypothesis, I feel it is quite likely that fallout plutonium from nuclear weapons tests has caused some cancer cases in people but that, on an individual basis, these plutonium-induced cancers cannot be distinguished from the massively larger number of "naturally" occurring cancers. I also think that a statement of "no proven human cancers from plutonium" is based on very weak evidence and puts the plutonium advocate in a very vulnerable position. The long latency for most radiation-induced cancers is well known and, if one proven case appears, not only does it destroy the presumed evidence for absolute safety of plutonium, but it can be used as counterevidence to destroy the position of the plutonium advocate. I believe it is much better to derive realistic numerical evaluations of the plutonium risk and to continue procedures to minimize contamination of people.

Table 9−3 shows the provisional risk coefficients I derived for a population of mixed ages (Mays 1976). For example, based on the linear hypothesis, if 1,000,000 people each receive an average skele-

Table 9−2. ^{239}Pu-Induced Malignancies

Bone Sarcomas
Liver Cancers
Lung Carcinomas

Table 9–3. Provisional Risk Coefficients for ^{239}Pu in Man, Based on Average Dose to Organs

$$\text{Bone Risk} = \frac{200 \text{ Bone Sarcomas}}{10^6 \text{ Person} \cdot \text{Rad}}$$

$$\text{Liver Risk} = \frac{100 \text{ Liver Cancers}}{10^6 \text{ Person} \cdot \text{Rad}}$$

$$\text{Lung Risk} = \frac{200 \text{ Lung Cancers}}{10^6 \text{ Person} \cdot \text{Rad}}$$

tal dose of 1 rad from plutonium, then about 200 would be predicted to die from plutonium-induced bone sarcomas. The bone risk coefficient is based on German patients given bone-surface-seeking radium–224 in repeated injections, continued over long periods of time, for which the effectiveness should be roughly similar to that from the protracted irradiation from plutonium.

The lung risk coefficient is based on the neutron-irradiation of the people at Hiroshima.

The liver risk coefficient is based on the Thorotrast patients. But in retrospect, I should have disregarded those patients who died before the ten-year minimal latent period for liver cancer, and I should have estimated, based on present rates, the number of living Thorotrast patients who will probably die from liver cancer. With these corrections, the liver risk coefficient becomes about 300 liver cancers/10^6 person · rad of α-radiation to the liver.

Table 9–4 is the most important table I will show you. If a young adult man inhales 1 microcurie of plutonium–239 in particulate form (1 micrometer activity median aerodynamic diameter of a class "Y" material), and if it is metabolized according to the ICRP report on plutonium (Lindenbaum et al. 1972), then over a period of fifty years, the primary lung tissue (excluding the lymph nodes) would receive an average dose of 30 rad, the liver would receive an average dose of 40 rad, and the marrow-free skeleton would receive an average dose of 13 rad. Multiplying each of these doses by the corresponding risk coefficients, the calculated probabilities of developing plutonium-induced cancers over the lifespan would be about 0.3 percent for bone sarcoma, 0.4 percent (or 1.2 percent, using the revised risk coefficient) for liver cancer, and 0.6 percent for lung cancer. Stated differently, if 100 persons each inhaled 1 microcurie of particulate plutonium–239, I would now expect about two of them to die with some kind of plutonium-induced cancer, based on present information. B.L. Cohen (1977) derives a similar risk estimate of

Table 9-4.   Predicted Incidences from Inhalation of 1 μCi ^{239}Pu

$$\text{Bone Sarcomas} = [13 \text{ Rad}] \left[\frac{200 \text{ Bone Sarcomas}}{10^6 \text{ Person} \cdot \text{Rad}} \right] = 0.3\%$$

$$\text{Liver Cancers} = [40 \text{ Rad}] \left[\frac{100 \text{ Liver Cancers}}{10^6 \text{ Person} \cdot \text{Rad}} \right] = 0.4\%$$

$$\text{Lung Cancers} = [30 \text{ Rad}] \left[\frac{200 \text{ Lung Cancers}}{10^6 \text{ Person} \cdot \text{Rad}} \right] = 0.6\%$$

Total Incidence of Pu Induced Cancers $\simeq$ 1%

Table 9-5.

Caution:

Plutonium Intake
May be Hazardous
to your Health!

one cancer death for every 40 to 80 microcuries that is inhaled. The actual risk could be considerably greater or less than these estimates, and additional work is needed for increased reliability.

Table 9-5 gives the take-home message that plutonium intake may be hazardous to your health, and, therefore, it is very important that we continue to restrict the intake of plutonium by people. If the plutonium intake is properly controlled, the risk from plutonium-induced cancer should not be the major issue on the safety of nuclear power. I think that adequate safeguards to prevent the making of unauthorized nuclear bombs is a much more crucial issue.

SUMMARY

In conclusion, plutonium-239 is not the most toxic substance on earth, but it must be treated with respect and be carefully safe-guarded. Dr. Rowland's dial-painter data fit a dose-squared response for bone sarcomas at low doses, but other examples of α-particle-induced cancers exist where the dose response appears to be approximately linear or even concave downwards. In view of this evidence, I believe it is more realistic to assume a linear dose response at low doses from plutonium α-radiation. The inhalation of 1 microcurie of plutonium-239 by a young man is calculated to give him about a 2 percent chance of developing a plutonium-induced cancer during his remaining lifespan. Additional reliability is needed in risk evaluation.

REFERENCES

Buldakov, L. A., and E. R. Lyubchanskiy. 1970. "Experimental Basis for Maximum Allowable Load (MAL) of Plutonium-239 in the Human Organism, and the Maximum Allowable Concentration (MAC) of Plutonium-239 in Air at Work Locations." Translation ANL—TRANS—864. Argonne, Ill.: Argonne National Laboratory.

Cohen, B.L. 1977. "Hazards from Plutonium Toxicity." *Health Physics* 32: 359—79.

Finkel, M.P. 1971. Personal communication.

Finkel, M.P.; B.O. Biskis; and P.B. Jinkins. 1969. "Toxicity of Radium-226 in Mice." In *Radiation-Induced Cancer*, ed. A. Ericson, pp. 369—91. Vienna: International Atomic Energy Agency.

Jee, W.S.S. 1978. Injection Tables, Research in Radiobiology. Report COO—119—253. University of Utah.

Kaplan, E.L., and P. Meier. 1958. "Nonparametric Estimation from Incomplete Observations." *Journal of the American Statistical Organization* 53: 457—81.

Lindenbaum, A., et al. 1972. *Metabolism of Compounds of Plutonium and Other Actinides*. International Commission on Radiological Protection, Publication 19, Oxford, England: Pergamon Press.

Luz, A., et al. 1976. "Estimation of Tumour Risk at Low Dose from Experimental Results after Incorporation of Short-Lived Bone-Seeking Alpha-Emitters [224]Ra and [227]Th in Mice." In *Biological and Environmental Effects of Low-Level Radiation*, vol. 2, ed. M. Lewis, pp. 171—81. Vienna: International Atomic Energy Agency.

Marshall, J.H., and P.G. Groer. 1977. "Theory of the Induction of Bone Cancer by Radiation. II. A Possible Low-Lying Linear Component in the Induction of Bone Cancer by Alpha Radiation." Report ANL—77—65, pt. II. Argonne, Ill.: Argonne National Laboratory.

Mays, C.W. 1978. "Addendum [to the F. Schales paper] Risk to Bone from Present [224]Ra Therapy." In *Biological Effects of [224]Ra*, ed. W.A. Müller and H.G. Ebert, pp. 37—43. Boston: Martinus Nijhoff Medical Division.

Mays, C.W. 1976. "Estimated Risk from [239]Pu to Human Bone, Liver, and Lung." In *Biological and Environmental Effects of Low-Level Radiation*, vol. II, ed. M. Lewis, pp. 373—84. Vienna: International Atomic Energy Agency.

Mays, C.W., and R.D. Lloyd. 1972. "Bone Sarcoma Incidence versus Alpha Particle Dose." In *Radiobiology of Plutonium*, ed. B.J. Stover and W.S.S. Jee, pp. 409—30. Salt Lake City: J.W. Press.

Mays, C.W.; H. Spiess; and A. Gerspach. 1978. "Skeletal Effects Following [224]Ra Injections into Humans." *Health Physics* 35: 83—90.

Mays, C.W., et al. 1976. "Estimated Risk to Human Bone from [239]Pu." In *Health Effects of Plutonium and Radium*, ed. W.S.S. Jee, pp. 343—62. Salt Lake City: J.W. Press.

Müller, W.A., et al. 1978. "Late Effects after Incorporation of the Short Lived α-emitters [224]Ra and [227]Th in Mice." *Health Physics* 35: 33—55.

Rossi, H.H., and C.W. Mays. 1978. "Leukemia Risk from Neutrons." *Health Physics* 34: 353–60.

Rowland, R.E.; A.F. Stehney; and H.F. Lucas, Jr. 1978. "Dose-Response Relationships for Female Radium Dial Workers." *Radiation Research* 76: 368–83.

Schales, F. 1978. "Problems and Results of a New Follow Up Study—[224]Ra in Adult Ankylosing Spondylitis Patients." In *Biological Effects of [224]Ra*, ed. W.A. Müller and H.G. Ebert, pp. 30–36. Boston: Martinus Nijhoff Medical Division.

Spiess, H., and C.W. Mays. 1970. "Bone Cancers Induced by [224]Ra (ThX) in Children and Adults." *Health Physics* 19: 713–29.

_____. 1971. "Addendum." *Health Physics* 20: 543–545.

_____. 1973. "Protraction Effect on Bone Sarcoma Induction of [224]Ra in Children and Adults." In *Radionuclide Carcinogenesis*, ed. C.L. Sanders, et al., pp. 437–50. AEC Symposium 29, CONF–720505. Springfield, Va.: National Technical Information Service.

Svec, J.; E. Kunz; and V. Placek. 1976. "Lung Cancer in Uranium Miners and Long-Term Exposure to Radon Daughter Products." *Health Physics* 30: 433–37.

Thurman, G.B., et al. 1971. Growth Dynamics of Beagle Osteosarcomas. *Growth* 35: 119–25.

Wick, R. 1978. Personal communication, June 8.

Discussion

Bernie Cohen: In case people got the impression from Dr. Rowland's talk that the Mancuso, Stewart, Kneale paper is accepted, I have the following comment: the speaker said that there were more cancers than expected at Hanford, and that is not true; among the people who died, there were more cancers than expected, but among all the exposed workers, there were fewer cancers than expected. At least fifteen critiques have come out against the Mancuso paper, and its thesis has been rejected by the ICRP, by the staff of the Nuclear Regulatory Commission, and by the United Kingdom National Radiological Protection Board.

Paul Kenney: To Dr. Mays: given the size of the error bars on your data points, I do not see any way of distinguishing between upward concave, linear, and so on. Is there a better chi square fit for the upward concave, rather than the linear or the others, and if so, by how much?

Charles Mays: There is so much scatter among the error bars, that one frequently cannot distinguish among these possibilities. But certainly, in Miriam Finkel's mouse data, the error bars are so tight, there is no question that this is a linear dose response.

Alvin Weinberg: I would like to ask Dr. Mays whether in the experiments that he has reported, there have been corrections for competing risks.

Dr. Mays: No, the corrections have not been made because the skeleton was the main target organ, and nearly all of the deaths caused by radiation were osteosarcomas. However, with total body irradiation, many cancers are induced in various organs, and then one should make corrections for competing risks.

Marc Roberts: Dr. Mays, you said that the ultimate mortality risk was 1 percent over fifty years from inhaling 1 microcurie of plutonium–239. What would be the normal mortality risk over fifty years for a twenty-five-year-old male, and so how much of an elevation is involved in the cancer from plutonium inhalation?

Dr. Mays: The risk of dying for a twenty-five-year-old male in the next fifty years is about 50 percent, but almost 100 percent by age 100. The chances of his dying from cancer during his lifespan, depending on his smoking habits, could be in the range of 15 percent up to 25 percent. An acceptable dose should be one that gives an acceptable risk. What is an acceptable risk? This is very hard to define because it depends on the importance you attach to a preventable cancer. As conditions in occupations get better and better, I think what is considered an acceptable risk is going to get progressively lower and lower. Personally, I would not like to see the people in my laboratory subjected to an occupational risk of 1 percent. If the risk were as low as 0.01 percent or perhaps even 0.1 percent, if this risk were unavoidable, it might be considered permissible. As we get more civilized we are going to lower our permissible limits for all occupationally induced diseases.

Mike Momeni: For Dr. Mays: experimental data, for example our data (Momeni 1979), indicate that a decrease in the level of radiation exposure induces an increase in the median observed mortality age (latency) and a broadening of the mortality distribution with respect to time. An extrapolation of cumulative incidence for any tumor observed at higher dose rates to levels within a few fractions of background radiation, based on any shape of dose effect curve, must incorporate the effect of shrinking population size at older ages from natural mortality with the decreasing exposure levels. This is supported by the table shown by Dr. Rowland. In experimental animal conditions, as the cumulative dose decreases, the comparative risk of natural mortality over all other causes will make the observed incidence in an animal obscure, difficult to see, and very difficult in any large animal experimental conditions. How can we, in work with the limited number of experimental animals available to us, such as beagles, correct for these low events? Linear extrapolation of radiation-induced effects to very low radiation levels, such as exposure from a single photon or alpha radiation, may indicate a potential late effect. I believe it is necessary to incorporate finite human lifespan and competition from naturally induced cancers within any extrapolation and radiation-induced risk analysis.

Dr. Mays: First of all, I would like to comment that Dr. Momeni did a lot of work on ^{90}Sr when he was at Davis, and in that case, the long range of the beta particles puts the young animal at risk, not only from bone cancer, but also from leukemia. In that type of analysis, it certainly was important that the correction for competing risks be made. My own view is that we cannot solve this problem by a purely experimental approach. The national budget would not buy enough dog or mouse food. On the other hand, I doubt that we would be able to solve this problem by a purely theoretical approach either, because we do not know enough yet about the mechanisms by which radiation induces cancer. I think what is needed is a healthy mixture of the theoretical approach, such as John Marshall has used so well here at Argonne, and work that the experimentalists do. We should try to fit together as many different lines of evidence as we can. This is more important than taking one experiment and doing exhaustive statistical tests on it, because that one experiment may not represent the generalization for all types of cancer.

Warren Sinclair: I would like to make two comments, one to Dr. Rowland and one to Dr. Mays. I certainly endorse Dr. Rowland's plea about comparative risks. I would like to point out that there is already one document in the literature (ICRP 27) that attempts to do exactly this. It is only a beginning attempt, but it does establish that current average occupational levels are in the ball park of safe industries, rather than harmful ones, which is comforting. But the attempt does exist, and I think there are going to be more of them.

To Dr. Mays: I am very impressed with the amount of dose effect information that you have presented. I would like to point out that they are all alpha-emitters, which of course, is very proper—the subject of this symposium is "Plutonium as a Test Case"—but the dose effect curves of X rays and gamma rays are also of great interest to us. I wonder if you care to comment on the shape of those?

Dr. Mays: Yes, I mentioned this very briefly in the first figure that I showed (Figure 9—1). The sigmoid dose response is the kind of curve that typically results from sparsely ionizing radiation. The

146 Decision Criteria for Health, Safety, and Environment

most common radiation that is encountered is probably medical
X rays, and the X rays give a diffuse distribution of ionizations. If a
single ionization occurs, isolated from others, there is a much greater
probability that the damage can be repaired, than if two lesions
occur close to each other at about the same time. So, then, one
would expect in the very low dose of X rays to have a very small
effect per unit dose, because it is highly unlikely that two points of
damage from the sparsely ionizating radiation would occur close
together. On the other hand, with densely ionizing radiation, such as
alpha particles, you have a chance of having many damage points
close together along the same track, and it is more plausible that the
alpha dose response should contain a stronger linear component than
does the beta, X-ray or gamma-ray dose response.

Gene Rochlin: I wish that one of the people working in this field
who holds the opposite point of view about Pu were here to speak. I
am going to speak out on their behalf, unfortunately in a field about
which I know very little directly, although a medical student worked
with me this summer trying to learn some of the other aspects of
plutonium fuel cycles. As far as the radium curves are concerned,
what Dr. Rowland showed was that if the sigmoid curve is the same
as for plutonium and if 3×10^{-5} is the correct figure for uptake
through the intestine, there would be effects. Now, there was recently a paper in *Science* that proposed that for chlorinated water
that figure might be several orders of magnitude off, in which case
the predicted number of cancers would be much higher. There are
also questions about chelated forms and organic complexes. Finally,
one has to question the data for plutonium in man, which is the
problem that we talk about. Fortunately, thanks to the experience
with the radium dial painters, there was very little exposure of
human beings to many other things, so the data are very poor. We
are glad of that, I think, although in the back of our minds we wish
we had better data, until you think of the consequences. You cannot tell, I believe, within several orders of magnitude, whether the
model used for cancer induction in man by plutonium is or is not
correct. One can screen out the most extreme models of the Goff-
man/Tamplin type, for it appears that they would have predicted
enough cancers so that some would have been seen. But even at one
order of magnitude less than their predictions, which is several orders
greater than presently accepted standards, there is not enough data
to say "yes" or "no." It is necessary to take a little caution on this
data.

With regard to commercial fuel cycles, there are two isotopes to which there is potential exposure, americium and curium, with which we have very little experience and for which our data are extremely sparse. Finally, I should point out that the radium that was used by the dial painters was measured in grams. Maybe there were kilograms of radium used altogether, but I doubt it at the price; what we're talking about with the plutonium fuel cycles are amounts of material being handled up into the thousands and hundreds of thousands of kilograms. The potential—I'm not saying that exposure must happen—but the potential reservoir, the potential amount of exposure, the potential consequences of an error are much larger than they have been historically. So I think people want to proceed with a bit of caution. What I have not seen and would like to see is a sensitivity check that says, "Suppose we are wrong by one order of magnitude; suppose we are wrong by two orders of magnitude?" When I start seeing that screening, I feel more comfortable because I can say, "Well, O.K., you've got a good margin against this uncertainty."

Robert Rowland: The article you referred to in *Science* (Larson and Oldham 1978) came from my laboratory. The work was done here, so we are fully aware that chlorinated water accelerates the transfer of soluble plutonium into an oxidation state that is readily absorbed through the gastrointestinal tract. Plutonium dioxide, as used in my example as a hypothetical dial paint, is not soluble, and that is why I made the statement at the end that "it depends very strongly on the form that it's in . . . ." That is one reason there is so much confusion on this whole subject of toxicity.

I think I can reassure you on another point. We now have under study a large group of workers from a plant in which they were exposed to a dust containing thorium. Many must have inhaled the dust, for we detect thorium in their lungs, or somewhere in their chest region. We have, at this moment, not a good measure of the total quantity in their lungs, but we do know that many contain enough thorium to receive approximately the same lung dose as would be delivered by the so-called permissible lung burden of plutonium. Work at this plant started in about 1933 and ended in 1973, so we have about forty years of experience from which to draw conclusions. It is far too early to give you the results, but I can reassure you to the extent that there is no epidemic of lung cancer. And that, in a sense, is giving you what you asked for. We may not know the numbers precisely, but we do know that alpha irradiation of the human lung is not one hundred times more dangerous than we predict, probably not ten times, and it might be even less dangerous than we predict.

Dr. Mays: I would like to say something about the "hot particle" hypothesis. It perhaps was unfortunate that when Tamplin and Cochran came out with this model of the "hot particle" hypothesis, they introduced a factor of 10^5. They claimed plutonium was 100,000 times more toxic than generally believed. Now, if they had said a factor of ten, it might have been credible, and it might have got the standards reduced by some factor, say, such as ten. But to overstate the case that badly, I think, put the whole argument in a bit of disrepute. Now, you ask about calculation; I can refer you to a document: *A Radiobiological Assessment of the Spatial Distribution of Radiation Dose from Inhaled Plutonium*, authored by Bill Bair, who is manager of Health Sciences at Battelle Northwest; Chet Richmond, who is currently Associate Director at Oak Ridge; and Bruce Wacholz, who is working with the Department of Energy (Bair, Richmond, and Wacholz 1974). They took the values of the Los Alamos cases that were exposed back when the Nagasaki bomb was built, and from this they calculated that if the Tamplin/Cochran "hot-particle" hypothesis were right, then among the twenty-five people who were most heavily exposed there should have been 5000 lung cancers (an average of 200 cancers per person). So far, there have been none. So we do have some human information that allows one to rule out absurd estimates.

Another very good example mentioned in the written version of Dr. Rowland's paper is the study of the thorium workers. But here is a case of an alpha-emitter in a human lung, and the fact that there has not been a dramatic incidence of lung cancer is telling us something. Maybe it is not very spectacular, like a factor 10^5 times more hazardous than widely believed, but at least it tells us that the estimates are not so terribly wrong. At least, we are not making gross underestimates of what the actual risk is. Please don't think that Dr. Rowland or I are trying to make any cover-up and pretend that plutonium is absolutely safe, that you can eat it by the spatula, provided it's not in valence +6! We are simply trying to find out what the risk is, because we are trying to let people who have decisions to make on energy alternatives know whether the carcinogenic issue is a major issue or whether it is only a minor issue. While it may not be very spectacular for us to tell you that the carcinogenic risk does not seem to be as important as most of the other considerations, I think it is something very important for the decisionmaker to have in mind: that we are looking at the toxicity of this substance; that indeed, it *is* potentially very hazardous, that is why it *must* be contained, but it is not as bad as some people have predicted.

Dr. Rochlin: No, I understand that, and I also understand that the Goffman/Tamplin/Cochran models multiply by several factors of ten, from four or five different mechanisms; it is possible that if they had treated those separately, they would have made a better case than by insisting that all of them were right. What I worry about is the kind of statement that I run into continuously; that is: "There has never been a known death due to the commercial nuclear industry." I am. not saying that either of you said that, but I think that people should be aware what the potential risks are.

REFERENCES

Bair, Bill; Chet Richmond; and Bruce Wacholz. 1974. *A Radiobiological Assessment of the Spatial Distribution of Radiation Dose from Inhaled Plutonium.* U.S. AEC Report WASH—1320. Washington, D.C.: U.S. Atomic Energy Commission.

International Commission on Radiological Protection (ICRP). n.d. Publication 27.

Larson, R.P., and R.D. Oldham. 1978. "Plutonium in Drinking Water: Effects in Chlorination in Its Maximum Permissible Concentration." *Science* 201: 1008—1009.

Momeni, M.H. 1979. "Competitive Radiation-Induced Carcinogenesis: An Analysis of Data from Beagle Dogs Exposed to [226]Ra and [90]Sr." *Health Physics* 36, no. 3: 295—310.

Session III
Decision Criteria for Nonproliferation Requirements

The proliferation issue has been of decisive importance in forming opinions on the possible use of plutonium as a fuel for nuclear power plants. Plutonium is produced in nuclear power reactors in a form that is not easily accessible, for example, in the fuel elements. Since it is a valuable nuclear fuel, it was assumed until recently that the spent fuel elements would be reprocessed to extract the plutonium as well as the residual fissionable uranium. However, the conventional reprocessing procedure is vulnerable to diversion of weapons-grade plutonium. Therefore, plans to construct reprocessing plants have been delayed or cancelled in the United States in order to discourage their construction elsewhere. Furthermore, plans to deploy the fast breeder reactor, which is intended to produce plutonium and use it as a fuel, have also been delayed. Since the known supplies of high grade uranium ore are limited, it is generally believed in the nuclear industry that the breeder will be required to maintain a nuclear power option for the future. Various alternatives to the fast breeder, presumably less subject to the proliferation hazard, have been proposed. This session is concerned with technical evaluation of these alternatives and with the political and economic implications, both national and international, of technical and institutional nonproliferation measures.

Session III

Opening Remarks by

Chairman: *William Sewell*

Our topic for discussion tonight is: "Non-Proliferation Criteria for Assessing Civilian Nuclear Technologies." I will not trouble you with introductory remarks; this is a topic about which I know little, and I am frank to admit it. I feel much like one of my students must have felt back in the mid–thirties when I was a young instructor at the University of Minnesota.

Minnesota had a National Championship football team, and it happened I had several members of the football team in my introductory course in Sociology. Among them was a young man who played guard, and who made the "Big Ten" but was flunking the course. Another was an All-American end, who was a Phi Beta Kappa, by the name of Ray King. Some of you may remember him, because he made All-American for two straight years. The coaches used to call me up and ask me what to do about this guard who was failing the course. I told them, "Get him to come to class." One day, he finally made it to class. He came around and said to me, "Bernie Berman wants me to work harder and pass this course, and I'd like to know how."

"Well," I said, "why don't you model yourself after Ray King? Whenever there's a game away, Ray comes to class, has a taxi waiting outside. When the class is over, gets into the taxi, goes to the Railroad Station (in those days they went by train), but first, he always asks me for the assignments. Monday afternoon he's in class, turns in his work and is always well prepared. Instead of that, what you do, whenever there's a game on Saturday, is to leave sometime on Thursday. You're too tired to come in Monday, and possibly you come in on Wednesday. Just pattern yourself after Ray King."

He just looked at me and said, "Mr. Sewell, don't you understand, dem ends, dey live in a different woild!"

Well, these experts on nuclear technology live in a different world than I.

✳ *Chapter 10*

Technical Considerations in Decisions on Plutonium Use

Charles E. Till

In the nonproliferation initiatives taken by the United States in late 1976 and early 1977 and since followed up through various implementation measures, attention was focused on plutonium, and more particularly, on its reprocessing, recycle, and reuse in civilian power reactors. The thrust of the policy is to defer its use, preferably indefinitely. It is an understatement to say that the policy has been controversial. To understand the reasons, a little technical background material will be required.

The raw materials from which fission weapons manufacture is possible are the two fissile isotopes of uranium, uranium–232 and uranium–235, and the two fissile isotopes of plutonium, plutonium–239 and plutonium–241. Uranium–235 is the only naturally occurring isotope. It is present in mined uranium in small quantity, approximately 0.7 percent of the total. The others are created in reactors, in present-generation power reactors, more or less as by-products of the power production process. But they are valuable by-products, for they can be recovered from the burned fuel and used to fuel further reactors.

All present-day commercial nuclear power generation is done using uranium–235 as the basic fuel, and indeed all nuclear power production must start with uranium–235 as the only fissile isotope found in nature. Only two reactor types are available commercially: the light water reactor (LWR) developed in the United States and since widely disseminated abroad, and the heavy water reactor (HWR or CANDU) developed in Canada. There are a few heavy water reactors around the world; the LWR is by far preeminent and is likely to remain so. In the LWR, enrichment facilities are required to increase the fissile uranium–235 content to about 3 percent in the fresh fuel. The CANDU reactor, on the other hand, can operate on natural uranium (without enrichment) and, in fact, it invariably does. In both, plutonium is produced during operation by neutron capture in the major isotope of uranium–238, rather more in the CANDU reactor than in the LWR, but in both in amounts that result in plutonium contents in the spent fuel discharge for 1000–MWe plants of a few hundred kilograms per year. Also in both, a substantial fraction of the plutonium created is burned in place, and in so doing, it produces about one-third of the total power of the reactor. However, once the uranium–235 content has burned down to the point where it can no longer sustain reactor operation, the spent fuel is simply removed and stored. For this reason, the present-day fueling procedure is called once-through cycle.

Uranium–233, which I will come back to later (for it is the basis of most of what are currently termed "alternate cycles"), is produced in an analogous way from the fertile material thorium–232, the naturally occurring isotope of thorium. That is to say, if natural thorium were substituted for uranium–238 in power reactors, the fissile material produced would be uranium–233.

All four of these fissile isotopes are capable either of power production or, in their purified forms, weapons manufacture. That is all there are, just these four. In the present-day, once-through cycle, the basic fuel is mined uranium, or, rather, the uranium–235 constituent from the mined uranium, with some bonus in power production from the plutonium created and burned in place. Basically, therefore, the once-through cycle is a process of mining and burning the naturally occurring fissile isotope of uranium. It is here that the concern about shortages of uranium, extent of uranium reserves, and so on, arises. For in this system, the amount of nuclear power generation possible depends more or less directly on the amounts of uranium producible. While there is much debate on the extent of the uranium resources, both in this country and internationally, by no one's estimate are the producible amounts large in comparison to estimated world needs for electrical energy for many decades in the future or for the portion that may be required from nuclear.

The technical means to alleviate this situation is to make better use, or further use, of the artificially created fissile isotopes, plutonium or uranium–233. But it is here that the perceived link between civilian power generation and nuclear weapons proliferation comes in. For all known practical ways of utilizing the other three fissile isotopes effectively involve chemical reprocessing of the spent fuel and reconstitution of the recovered fissile material into new fuel elements. The reprocessing step provides the ability to produce more or less pure fissile material. In the case of the uranium–plutonium cycle, that material is plutonium, not necessarily in the best chemical or isotopic form for weapons purposes, but nevertheless one step closer than allowing it to remain in the spent fuel. Thus, the further use of plutonium in nuclear power generation, through reprocessing and reuse, is seen as easing the way to weapons acquisition. It is reasoning along these lines that underlies the present policies of the United States regarding nuclear fuels.

The importance of limiting the spread of nuclear weapons I take as obvious; certainly in this country there would be nearly unanimous agreement on this point. What is not so obvious is the directness of the link between the fuel cycle for civilian nuclear power and weapons proliferation, as well as the importance of this link relative to other means of acquiring nuclear weapons. Furthermore, both in this country and abroad, there is great concern about what all this implies, in what would have to be foregone. The banning or indefinite deferral of plutonium reprocessing and reuse means a sacrifice in energy production capability. To understand the extent of this sacrifice, it will be necessary to examine some more technical details.

THE IMPLICATIONS OF THE FUNDAMENTAL PROPERTIES OF NUCLIDES ON POSSIBILITIES FOR NUCLEAR POWER GENERATION

The four fissile isotopes, while they have broadly similar nuclear properties, differ in important detail in their basic nuclear properties. Some of these differences are absolutely crucial to the performance that can be expected of them in power-generating reactors and, in turn, to the magnitude of the power that can be generated from a given amount of natural uranium. To a somewhat lesser, but still important, extent the same is true of their two fertile predecessors, thorium–232 and uranium–238. It should be stressed that these differences are fundamental and unchangeable by modifications in nuclear reactor design. All that nuclear reactor design can do is to take advantage, to a greater or lesser degree, of the properties that these isotopes inherently possess. Insofar as fuel savings are concerned, the important factors in nuclear reactor design are its fissile material requirements. The basic principles are shown in Table 10–1.

The most important factor in extending the amount of energy that can be produced from a given amount of natural uranium is the degree to which the fertile isotope can be converted to fissile material during reactor operations. This factor differs markedly between various possible practical reactor designs. Referred to as the *conversion ratio* in thermal reactors (essentially present-day commercial reactor types), and as the *breeding ratio* in fast breeder reactors, the definition of this factor and the nuclear properties determining it are shown in Table 10−2.

Table 10−1.   Basic Reactor Fissile Material Requirements

- A substantial amount of fissile material is required at all times to keep the reactor operating (for criticality). This is called the *inventory*.
- A certain amount of this is burned to produce power.
- In the power-production process, neutron capture in the fertile material produces new fissile material.
- If the amount of new fissile material produced is less than the amount burned, the reactor is called a *converter reactor*, and the reactor will have yearly *fissile requirements* that must be provided to it.
- If the amount of new fissile material produced is more than the amount burned, the reactor is a *breeder*, and it will have a yearly *fissile gain* that can be used either to provide the yearly fissile requirements of a converter reactor, or after accumulation, to provide the necessary inventory to start up another breeder.
- The index of the net production of fissile material is called the *breeding ratio* (in breeder reactors) or the *conversion ratio* (in converters).

Table 10−2.   Relation between Fissile Material Requirements and Basic Nuclear Properties

Breeding Ratio or Conversion Ratio

$$= \frac{\text{Fissile material created during reactor operation}}{\text{Fissile material burned during reactor operation}}$$

$$= \eta - 1 + \epsilon - \text{Losses} \quad \text{(dependent upon basic nuclear properties and particular reactor design)}$$

where

$$\eta = \frac{\text{Neutrons produced by fissile isotope}}{\text{Neutrons absorbed in fissile isotope}} \quad \text{(a property of fissile isotope)}$$

$$\epsilon = \frac{\text{Neutrons produced by fertile isotope}}{\text{Neutron absorbed in fissile isotope}} \quad \text{(a property of fertile isotope)}$$

and,

losses are due to capture of neutrons in necessary structure and control materials in the reactor and to leakage of neutrons from the reactor.

These properties differ from isotope to isotope, and very importantly, they change markedly as the neutron energy is changed. In a fast neutron spectrum, they are considerably different than they are in a thermal neutron spectrum. The effect of these differences is shown dramatically in Table 10−3. There are several aspects to note. First, plutonium/uranium-238 makes a very superior fast reactor fuel. The main fuel forms that have had work done on them as candidates for fast breeder fueling are: oxide, carbide, and metal—oxide involved in by far the most work. The breeding ratios range from 1.63 for metallic fuel form, to 1.33 for oxide fuel. The faster the neutrons, or what is the same thing, the harder the spectrum, the better the breeding performance. The metallic fuel form allows the hardest spectrum, and therefore, it has the best potential breeding performance. The presence of the light oxygen atoms in the oxide fuel softens the spectrum and lowers breeding performance. Carbide fuel is intermediate between the two. Secondly, the uranium-233/thorium cycle has considerably poorer performance for all fuel forms in the fast breeder, and it does not vary significantly with hardness of spectrum. Again, these are the result of fundamental nuclear properties, inherent in the fertile and fissile isotopes involved.

Equally importantly, the situation is reversed in a thermal reactor spectrum, where plutonium/uranium-238 is not an exceptional thermal reactor fuel, comparable to uranium-235/uranium-238, and uranium-233 is considerably the best of the three. No thermal reactor cycle, however, can achieve conversion or breeding ratios significantly above unity, although designs using the uranium-233/thorium cycle may approach unity, but even then only with a considerable penalty in the inventory of fissile material required to maintain operation.

Conversion and breeding ratios are not academic indices—they go a long way toward determining the efficiency of fuel utilization of the various reactor types. As a simple way of thinking about this, any reactor producing 1000 MWe of electricity, over a year on a normal duty cycle, will burn about 1000 kg of fissile material. Each kilogram burned will in turn create fissile material in an amount in kilograms about equal to the breeding or conversion ratio. For breeding ratios greater than unity, there will be a net gain; for less than unity a net loss. Accurate calculations of the yearly fissile requirements for the various reactor types are given in Table 10−4 for typical 1000-MW electric reactor designs operating at 75 percent capacity factor. Again, the point to note is the superiority of plutonium-239 in the fast spectrum, and its relative inferiority to uranium-233 in thermal spectrum.

Table 10–3. Typical Examples for 100 MWe Reactor Types

Fuel Cycle	Fast Breeder Reactor (LMFBR)				Thermal Converter Reactor (LWR)		
	Pu/^{238}U			^{233}U/Th	^{235}U/U	Pu/U Oxide	^{233}U/Th
Fuel Form	Metal	Carbide	Oxide	Metal, Carbide or Oxide			
η	2.45	2.35	2.28	2.28 ± 0.01	1.92	1.95	2.18
ϵ	0.51	0.43	0.36	0.09 ± 0.01	0.10	0.11	0.03
Losses	0.33	0.28	0.31	0.26 ± 0.01	0.44	0.38	0.47
BR or CR	1.63	1.50	1.33	1.11 ± 0.01	0.58	0.68	0.74

Table 10-4. Fissile Material Requirements for Typical 1000 MWe Reactors *(75% Capacity Factor)*

	Fast Breeder Reactor (LMFBR) Yearly Fissile Gain[a]		
U/Pu Metal	*U/Pu Carbide*	*U/Pu Oxide*	*Th/^{233}U*
420 kg Pu	365 kg Pu	260 kg Pu	~ 75 kg ^{233}U
	Initial Inventory Required		
2500 kg Pu	2600 kg Pu	3150 kg Pu	~ 3000 kg ^{233}U

	Thermal Converter Reactor (LWR) Yearly Fissile Requirements	
Th/^{233}U	*^{238}U/^{235}U*	*^{238}U/Pu*
	800 kg ^{235}U Feed −165 kg Pu Produced −210 kg ^{235}U/U in Spent Fuel	
235 kg ^{233}U	425 kg	380 kg Pu
	Initial Inventory Required	
1200 kg ^{233}U	1800 kg ^{235}U	1700 kg Pu

[a] "Gain" equals the amount of fissile material produced less the amount consumed.

The foregoing considerations provide the basis for understanding what is at the heart of the plutonium debate. If the fundamental nuclear properties of the plutonium and uranium fissile isotopes were similar, or, put another way, if thermal reactors only were possible, much of the controversy would evaporate. For as can be seen from Table 10-4, the net plutonium production in an LWR is about 165 kg per year, and the yearly feed requirement is about 800 kg of uranium-235. By recycling, about 40 percent of the annual feed requirement can be recovered, but half of this is in the recovered uranium-235. The net fuel savings in reprocessing and reusing the plutonium itself amounts only to some 20 percent. Although this is a useful saving, it almost certainly can be achieved by other, unrelated improvements in the LWR fuel cycle, and in all likelihood would not justify the investments in the recycle technology required. This last is a debatable point and would certainly not be agreed to by everyone. When one turns to what is possible in a fast breeder,

Table 10-5. Approximate Energy Generation Implications. Reactor
Lifetime Usually Considered to be 30 Years

1. For LWR once-through cycle:
    - lifetime ^{235}U requirement   1800 kg initial inventory =   1800

      29 yearly reloads (at

      800 kg/year)   = 23,200

      Total Requirement   25,000 kg

    - one ton of mined U_3O_8 yields 4 kg of ^{235}U
      (at 0.2% tails)

    - ∴ one 1000 MWe reactor requires 25,000 ÷ 4
      ≈ 6000 tons of U_3O_8
      and each million tons of mined U_3O_8 will allow 170 GWe for 30 years

2. For fast breeder reactors
    - the plutonium production from each LWR is:
      165 kg/yr × 30 years = 5000 kg.

    - each oxide-fueled breeder requires 3150 kg for its initial core, plus about
      half of this for material in process outside the reactor, that is, about
      5000 kg.

    - so each LWR can start up one breeder, which will then sustain itself and
      after accumulation of its fissile gain start up further breeders.

    - in this way growth of the generating system is possible, without further
      mining requirements.

the situation changes dramatically. There the uranium–238/pluto-
nium cycle is unique. Its property of creating appreciably more fissile
material than it consumes holds out the promise of energy production without the need for concern about uranium supplies. And of
course, it was this that caused the fast breeder, almost without exception, to be the international choice for development of a next
generation of reactor types. A rough indication of the implications
of these considerations is given in Table 10–5.

AN EXAMPLE OF THE URANIUM
REQUIRED TO MEET WORLD NUCLEAR
ENERGY DEMAND USING VARIOUS
TYPES OF REACTORS[a]

Proliferation of nuclear weapons has meaning only in an international context. In discussions of trade-offs between increased risk of

[a] Excluding centrally planned economies.

proliferation and adequacy of energy production, international energy production capability is at the heart of the issue. Sacrifices of nuclear energy production capability by the United States may have foreign policy implications, positive or otherwise, but ultimately it is limitations on other nations—their capability for acquisition of nuclear weapons or their means of energy production—that are under discussion.

Present and future energy needs of individual nations and their corresponding national resources obviously differ widely. Individual nations' perceptions of the future availability of uranium resources, however, must all start with their assessments of aggregate world uranium resources and production capability and likely aggregate world demands on these resources. Concerns about individual national shares of resources and relative assurance of such shares would then follow. But the first-order considerations are the global aggregates.

To illustrate the effects on uranium consumption of choice of reactor type, I have taken the latest OECD Nuclear Energy Agency (1978) estimates of world nuclear electric demand and resource availability as a data base (excluding centrally planned economies).

The OECD-NEA estimates of future nuclear electric demand are shown in Figure 10–1. Two forecasts are given, the first labeled "present trend" and the other "accelerated" demand. The accelerated demand was postulated on a return to a post-oil embargo sense of urgency, which at present, at least, is being heavily discounted. I have taken the present trend case, extrapolated linearly beyond the year 2025, for my example.

The OECD-NEA cites resources using the NEA–IAEA classification scheme, in two categories—"reasonably assured" and "estimated additional" resources. These categories are defined as follows:

"Reasonably assured resources" refers to uranium that occurs in known mineral deposits of such size, grade and configuration that it could be recovered within the given production cost ranges, with currently proven mining and processing technology. Estimates of tonnage and grade are based on specific sample data and measurements of the deposit characteristics.

"Estimated additional resources" refers to uranium in addition to reasonable assured resources that are expected to occur, mostly on the basis of direct geological evidence, in extensions of well-explored deposits believed to exist along a well-defined geological trend with known deposits. (OECD Nuclear Energy Agency 1977:11)

Figure 10–1. International-Aggregate Nuclear Energy Demand Projections
(Data Source: OECD Nuclear Energy Agency 1978)

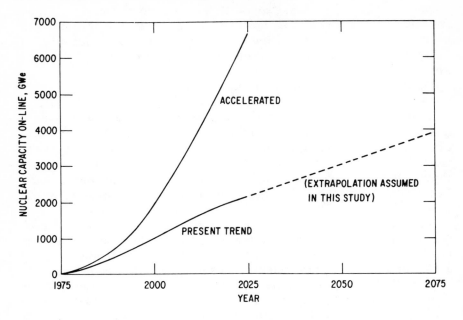

In terms of these categories, the OECD–NEA estimates for world-wide uranium resources at less than \$130/kg (\$50/lb) are given in Table 10–6.

A sizable fraction of these estimated resources is in the United States. In this connection, it is useful to note the correspondence between the OECD–NEA estimates and the current U.S. estimates. To begin with, the "reasonably assured" and "estimated additional" resources up to \$130/kg (about \$50/lb) are roughly equivalent to what the United States has usually referred to as "reserves" and "probable potential resources," respectively. The U.S. agencies responsible for these estimates also estimate additional potential resources, termed "possible" and "speculative," still at less than \$130/kg, that would increase the total U.S. resources to about 3.2 million tonnes (metric tons) of uranium. (In addition, efforts in the United States are now focused on estimating the available resources as a function of costs, with costs going beyond the limits just given.) As the inclusion of "possible" and "speculative" resources doubles in the U.S. resource base, it is reasonable to indicate the effect of at least doubling the international resource numbers as well.

Table 10-6. World Uranium Resource Position *(Thousand Tonnes U)*[a]

	Reasonable Assured Resources		Estimated Additional Resources		Total
	< $80/kg U[b] (< $30/lb U_3O_8)	$80-130/kg U ($30-50/lb U_3O_8)	< $80/kg U (< $30/lb U_3O_8)	$80-130/kg U ($30-50/lb U_3O_8)	
Australia	289	7	44	5	345
Canada	167	15	392	264	838
South Africa	306	42	34	38	420
United States	523	120	838	215	1696
Other African Countries	218	6	118	10	332
Western Europe	57	325	41	37	460
Countries not included above	87	29	44	16	176
Total (rounded) 10^3 tonnes U	1650	540	1510	590	4290
10^3 ST U_3O_8	2150	700	1760	770	5580

[a] Reproduced from OECD Nuclear Energy Agency (1978).
[b] Defined as "reserves."

Figure 10–2 shows the results, in terms of cumulative uranium consumption, of choice of reactor type in meeting the postulated "present trend" demand for the following four simplified reactor deployment scenarios:

- LWR once-through cycle only;
- LWR utilizing uranium and plutonium recycle;
- LWR once-through cycle followed by advanced converter reactor (ACR) (thorium/^{233}U cycle);
- LWR uranium recycle followed by the fast breeder reactor (LMFBR).

The advanced converter reactor was represented by reactors operating on a uranium–238/thorium cycle, with uranium–235 feed. For

Figure 10–2. Cumulative Uranium Consumption as a Function of Time for Various Deployment Options

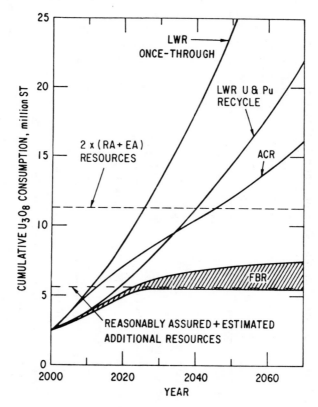

the LMFBR case, the range of performance characteristics is spanned by the three different fuel types. The introduction date for both ACRs and LMFBRs was assumed to be the year 2000, with an introduction rate of 150 GWe total during the first five-year period and 300 GWe total during the next five-year period. These clearly are very high introduction rates. At the very least, they imply that well before the year 2000 there must be much successful development work and deployment of at least several reactors of the commercial type, assumed to be introduced in quantity in the year 2000.

Figure 10−2 places the cumulative uranium consumption for the four scenarios against the background of the "reasonably assured" (RA) plus "estimated additional" (EA) resource level, and the arbitrarily doubled resource level. The continuous resource usage of converters, greater or lesser, depending on type, exhausts the RA plus EA resources soon after the year 2010 for any of the converter scenarios. Further, if 30-year forward requirements were accounted for, the point of complete commitment of RA plus EA resources would be reached by the year 2000. If these resources were all that were to be available, phase-out of nuclear power should therefore begin, for reactors constructed later than this would not have fuel supplies for their anticipated lifetimes.

Introduction of advanced converter reactors does not help, if they are to be introduced in significant numbers only on this time scale. By the date, probably the earliest realistic date, of the year 2000 for any such sizable ACR introduction, the LWRs constructed to that date would already have consumed the reference resources. Even allowing for further resources, it is worth noting that the ACR cumulative uranium consumption follows the LWR once-through curve for quite a time before it finally starts to drop below it. It is only after consumption of resources about double the reference value that the ACR shows any improvement over the LWR on recycle. This is due to the fact that advanced converter reactors require more uranium than LWRs during their start-up period. At equilibrium, the uranium requirements are reduced significantly, but for such savings to be substantial, the energy demand must not continue to grow.

For all breeder cases (the top of the shaded portion shows an oxide-fueled case, the bottom shows carbide or metal) the uranium consumption for this demand schedule begins to slow significantly around 2025. With the advanced breeder cycles no additional uranium is consumed after 2030, and the consumption leveled off just under the reference value of the resources. For the lower-gain breeders, the uranium consumption continues but eventually levels off when the breeder fissile doubling capability catches up with the energy growth.

In summary:

- For the smaller resource base, the incentive for the advanced converter reactor recycle is nonexistent, because the resource base is consumed mostly in the LWR once-through cycle before the advanced converter reactor can be deployed.
- For the large resource base, LWR recycle is about as resource-effective as the ACR.
- The fast breeder reactor energy supply potential is high in either case. In the smaller resource base case, advanced breeders using fuels other than oxide would actually meet the demand, essentially forever; in the large resource situation, any breeder would.

ALTERNATIVES TO PLUTONIUM:
THE "ALTERNATIVE CYCLES"

The presidential nuclear power policy statement of April 7, 1977, had as its third point, "that we will redirect funding of U.S. nuclear research and development programs to accelerate our research into alternative nuclear fuel cycles which do not involve direct access to materials usable in nuclear weapons." In the ensuing months, the precise definition of such alternative nuclear fuel cycles has not been made absolutely clear. Later policy statements have defined "alternative nuclear fuel cycles" as alternative to an economy based on the separation of pure plutonium or high enriched uranium. In the studies that were set underway by the policy announcements, however, alternative cycles have, in effect, been defined as being of two types. One type includes variants based on uranium–233/thorium cycles, and the second are variants of the current once-through cycle. The reason for this small number of alternatives is not hard to understand, if we refer back to the discussion of the basic nuclear properties given in Section 2. For if plutonium recycle is ruled out, the only two fissile isotopes remaining are uranium–235 and uranium–233. Uranium–235 is the basis for the once-through cycle. But uranium–233 is the bred fissile isotope in the thorium cycle and, as such, it is closely analogous to plutonium in the plutonium/uranium–238 cycle. The basic alternatives are therefore very limited.

The types of reactors that are implied by these alternate cycles also follow directly from the basic nuclear properties previously discussed. Thermal reactors are implied. This is not to say that fast reactors could not operate on a uranium–233/thorium cycle. In principle, of course, they could, but as we have seen in Section 2, their breeding properties would be sharply limited, and the incentive to undertake the effort to develop and deploy the fast breeder under these circumstances would be correspondingly limited. Moreover, a fast breeder uranium–233/thorium cycle would require reprocessing efforts of a kind very similar to those required for the plutonium/uranium–238 cycle. In effect, the only real change would be the substitution of uranium–233 for plutonium, but still in a very similar cycle; gain from the viewpoint of nonseparability of fissile material would certainly be small—in all likelihood nonexistent.

Once-through cycles are still less suited to a fast reactor. The higher initial enrichments of the fast breeder, compared to thermal reactors, and the correspondingly higher fissile contents discharged in spent fuel, give net fuel utilizations that are much worse than those for the present-day LWR, much less any improved version. Furthermore, it is difficult to see how any practical modifications to the fast breeder could change this appreciably. Mixes of reactor types that include some fast breeders and some thermal reactors operating on alternative cycles, termed "symbiotic cycles," are possible and have been the subject of some investigation. These I will come back to later, for they form one class of suggestion for balancing energy production requirements with nonproliferation goals. But if the alternate cycles alone are considered, it is almost certain that thermal reactors alone are involved.

The next thing that can be said with reasonable certainty is that even in thermal reactors, once-through cycles rule out thorium use. A once-through cycle implies uranium–235 fissile fueling, with thorium as the fertile element in place of uranium–238. The higher neutron absorption of thorium, again a fundamental nuclear property, requires higher initial enrichments to maintain reactor operation. In any calculation of the resulting fuel utilization in practical thermal reactor types, the resulting fuel utilization is invariably worse than for the corresponding uranium–235/uranium–238 once-through cycle. The reason is the same as the reason in the fast reactor case, albeit to a lesser degree. In both, the fissile contents in the discharged fuel remain high and too much fissile material is discarded in the spent fuel.

The next important fact that follows from these considerations is that the uranium–233/thorium, even in thermal reactors, depends on reprocessing the spent fuel for its potential gains in fuel utilization. With such reprocessing, substantial gains in fuel utilization over the current once-through cycle are possible. Net uranium–235 consumptions of one-half those of current once-through cycle probably are easily achievable and consumptions down to one-quarter those of the once-through cycle are possible, at least in principle. It is this cycle and the reactors in which such cycles would be used that form the class of reactors referred to in the previous section as advance converter reactors (ACRs). The reactors themselves are variants of current reactor types and differ in the main in the type of moderator selected. If heavy water, the ACR is a CANDU; if graphite, a high-temperature gas-cooled reactor (HTGR); and if ordinary water, an LWR. In all, particular design attention is directed to minimizing neutron losses, sometimes by very significant design changes, as in the case of the light water breeder reactor (LWBR). (The latter is something of a misnomer in that the maximum breeding of conversion ratio possible in this concept, is essentially unity.) In any event, it is in the latter, in minimizing neutron losses, that the design of the LWBR or any other ACR must accomplish its aim of improving conversion. Referring to the last column in Table 10−3, for example, the losses must be cut from the number shown there (0.47) to something less, and each such reduction increases the conversion ratio correspondingly. In so doing, the ACR strives for as high a conversion ratio as possible, consistent with other practical limitations.

The CANDU reactor, which was the basis for the ACR calculation shown in Table 10−7, probably represents something of an upper limit, because heavy water has the lowest intrinsic neutron absorption of any of the moderator types. But for it, and for all other ACRs as well, the requirement to minimize neutron losses implies relatively short burn-ups (to minimize losses in fission-product capture) and correspondingly high fuel discharge rates. This, in turn, increases the amount of reprocessing required, the cost, and the amount of fissile material always in process. Further, increasing the conversion ratio in ACRs also increases the required inventory. In

Table 10−7. Suggested Measures for Proliferation Resistance

Three Main Classes:

I. Fissile Material Form
   A.   Fissile/Fertile Mixture
   B.   Chemical Form
   C.   Isotopic Form
   D.   Associated Radiation

II. Administrative Measures: International and National Safeguards
   A.   Physical Protection
   B.   Materials Accountancy

III. Internationalization of Sensitive Facilities and Supplier Nation Arrangements all ACR cycles there is a trade-off between fuel utilization on one hand, and inventory and burn up on the other and, therefore, in the amount of reprocessing required annually. In all, reprocessing is fundamental. In this, they are basically similar to the fast breeder reactor and its fuel cycle. Discussions of possible improvements in the proliferation resistance provided by ACRs, therefore, tend to focus on other measures relating to proliferation resistance that they may help facilitate. This brings us to our next topic.

SUGGESTED MEASURES FOR INCREASING PROLIFERATION RESISTANCE: SOME TECHNICAL IMPLICATIONS

The measures that have been suggested to increase the proliferation resistance of the nuclear power fuel cycle, and that are included in most discussions of the subject, can be divided into the three main classes shown in Table 10−7. Most overall schemes include combinations of several of the items.

I. *Fissile material form*

A. *Fissile/fertile mixture* —implies the requirement that at no time is the fissile material to be separated from its associated fertile predecessor. Two main cases are normally discussed.

1. *The once-through cycle* —as the fuel, by definition, is never reprocessed, no facilities exist within the nuclear fuel cycle itself to separate plutonium from the spent fuel. Problems exist at both the front and back ends of the cycle. For LWRs, or in fact any converter reactor except CANDUs, enrichment facilities are required (and even CANDUs would benefit from enrichment). These facilities give capability for separation of pure fissile material (uranium-235). At the other end of the cycle, the spent fuel containing plutonium accumulates, and the burden is placed on arrangements for satisfactory retrieval (if in a sensible location), safeguarding and storage of continuously increasing amounts of such material. Initially highly radioactive, its activity decreases fairly rapidly as time passes, so that after a few years, it becomes fairly easily accessible.

2. *Coprocessing* —the international choice for reprocessing uses the Purex process, in which separation of the uranium/plutonium mixture from the highly radioactive fission products is followed by a partitioning step which separates plutonium from the uranium, and leaves a very clean, pure plutonium product. Modification of the chemistry of the process is possible to carry along a portion of the

uranium with the plutonium through the entire process. It is termed coprocessing and is probably feasible on a commercial scale, with further development work. Its conceptual problem is that it would be relatively easy to alter the process or produce a pure product, if that were desired.

    B. *Isotopic form*—implies the requirement that only isotopic mixtures of fissile and fertile would be allowed. That is to say, separation of pure fissile from the fissile-fertile mix could not be accomplished by normal chemical means, for the chemical species would be identical. This has been referred to as "isotopic denaturing." There are two such possibilities.

    1. *Uranium denaturing*—the main version of the isotopic denaturing idea. Fissile uranium–233 or uranium–235 are to remain mixed with relatively larger amounts of uranium–238 at all times. There are some conceptual problems. First, although certainly the case at the present time, it is not at all clear that isotopic separation (enrichment) capability will remain significantly more difficult to acquire than chemical separations capability. Second, for any cycle but the once-through cycle, that is, for any cycle that uses the artificially created fissile isotopes to extend energy-generation capability reprocessing remains an essential component of the cycle. Denaturant concentrations discussed in most studies involve 12 percent uranium–233 or 20 percent uranium–235 in total uranium, with the uranium itself then mixed with thorium. This mixture allows thorium to remain the main fertile species in the fuel, and the fuel remains suitable for ACR fueling. Typical proportions in an ACR fuel would be one part uranium–233, seven parts uranium–238, twenty-five parts thorium. Reactor operation produces uranium–233 from the thorium, plutonium from the uranium–238. Three-way separations are required in reprocessing. This appears to be technically feasible, but has not been demonstrated. The important point, however, is the required presence of the reprocessing step in any case.

    2. *Plutonium denaturing*—reactor-produced plutonium is always a mixture of the fissile isotopes, plutonium–239 and plutonium–241, and the fertile isotopes, plutonium–240 and plutonium–242. The process is plutonium–239 production from uranium–238, plutonium–240 from plutonium–239, plutonium–241 from plutonium–240, and plutonium–242 from plutonium–241, each step by successive neutron capture during reactor operation. The proportions of each are largely set by the reactor performance and fueling requirements, and in practical situations, the fissile fraction will

predominate. While higher concentrations of plutonium-240, in particular, make the material less desirable for weapons purposes, it appears to be impossible to rule out its usefulness on the basis of normal reactor plutonium-240 concentrations. A variant of this idea is to decrease the usability by attempting to increase the plutonium-238 content—the latter is a minor constituent of reactor-produced plutonium that has the property of relatively high heat generation. However, it is not produced in appreciable amounts in fast breeder fuel cycles, and as the great incentive for plutonium use derives from its use in the breeder, it is difficult to assign high importance to the idea.

C. *Associated radiation*—fuel cycle arrangements in which lethal levels of radiation are associated with the fuel have been suggested. Included in this category are the spent fuel in the once-through fuel cycle, at least for a period of years, and various design arrangements to make this the case in other fuel cycles. Among these are:

1. Partial-decontamination in the reprocessing step, bringing fission products along with fissile products.
2. Spiking, in which specially produced radioactive isotopes are added to the product.
3. Preirradiation of fresh fuel.

All increase costs, sometimes sharply, and difficulties and dangers to personnel in handling fuel are also increased.

Some significance is occasionally attached to chemical form also, with preference for ceramic forms over metals. The conceptual problem with this, and indeed, with all of the foregoing, is that if the technological sophistication and the will to manufacture nuclear weapons is assumed, the technological barrier presented by any of the above is not, and cannot be, very high. The basic question is whether constraints in this area are worth the price in difficulties in the energy generation. In combination with some or all of the remaining measures to be mentioned below, however, some of the foregoing measures may be useful.

II. *Administrative measures: international and national safeguards*

A. *Physical protection*—the guarding arrangements that control personnel entry and exit at sensitive facilities include personnel identification, monitors of various kinds, delay systems to slow down intrusion into a restricted area, and so on.

B. *Materials accountancy*—the assaying and counting procedures aimed at verifying that all fissile material is actually where it should be, and no diversion is taking place. The bulk of the IAEA safeguards activities are in this area.

The main technical implications of these measures are, first, that development, probably feasible, is required in instrumentation of various kinds to carry out these functions adequately, and second, possible incompatibilities between requirements in this area and suggestions with respect to fuel-form requirements must be recognized. An example of the latter is the incompatibility of high radiation level (I-C) with highly accurate assaying and identification instrumentation (II-B).

III. *Internationalization of sensitive facilities and supplier nation arrangements.* Implied in this are agreements between nations that reduce national freedom of action to utilize civilian nuclear programs to move toward weapons acquisition. Two main classes have been suggested.

A. *Nuclear supplier nation agreements*—restrict access to sensitive (generally separations-related) technology.

B. *Fuel cycle centers*—a concept with various forms, the basic notion of which is the concentration of the fuel supply facilities— enrichment, reprocessing, and possibly breeder reactors—in a limited number of locations, preferably under multi- or international auspices. Some forms have been called "secure area" deployment.

Now all of the foregoing have technological implications in the barrier that they may represent to weapons acquisition, and on the other hand, to the limitation that they represent to the ability of the civilian nuclear power programs to generate electrical power. It is fair to say that proliferation resistance assessments, insofar as they have gone to the present time, have tended to concentrate on differences between fuel cycles in the first area, that is, upon the fissile material form.

AN EXAMPLE OF A PROLIFERATION-RESISTANT CONCEPT: THE SECURED AREA SYMBIOTIC CYCLE

An illustrative scenario that incorporates many of the measures listed in the previous section is that of secured-area symbiotic cycles. In this concept, breeders utilizing fuels with chemically separable fissile materials (uranium/plutonium, or conceivably, uranium–233/thorium) would be located only in secured enclaves, along with the reprocessing and fuel fabrication facilities required both for the breeders and for converter-reactors (located anywhere) whose fueling requirements would be met by breeder-produced fuel. The concept is represented schematically in Figure 10−3. The fuel produced by the breeder for converter use is uranium-233, in turn supplied to the converter in isotopically denatured form. The converter spent fuel is returned to the secured-area, still highly radioactive, where it is reprocessed. The recovered uranium-233 is recycled back to the converter and the plutonium is fed to the breeder. The whole concept involves international or multinational control of the nuclear fuel cycle, a concept that can be traced back to earliest efforts to find ways to constrain the uses of nuclear energy to peaceful purposes.

In this concept, the breeder utilizes at least some thorium as fertile material. The impact on breeding of the replacement of fertile uranium-238 by fertile thorium-232 in fast breeder reactors always leads to a decrease in breeding ratio. The decrement in breeding ratio is not large if the thorium is confined to the blankets of the breeder, but it is substantial if thorium is used in the reactor core. In either case, the result of replacing uranium-238 by thorium is a replacement of bred plutonium by bred uranium-233. Increasing replacements of fertile uranium-238 by fertile thorium-232 in breeders leads to the following two effects:

1. Increasing amounts of uranium-233 are available for fueling larger numbers of dispersed converters (or possibly breeder reactors operating on a denatured uranium-233 fuel cycle).
2. Less plutonium is available, however, to continue to expand the number of breeder reactors which are the very "factories" in which the uranium-233 is generated.

Figure 10−3. Secured-Area System

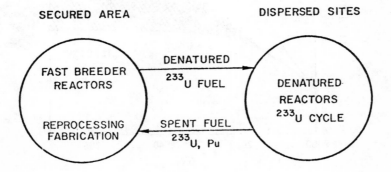

It is clear, then, that a trade-off exists between the rate at which the symbiotic system can grow to meet demand (effect 2) and the ratio of dispersed converter reactors to confined uranium-233 generating breeders (effect 1).

The effect on uranium consumption of varying the thorium content in the breeder for the same OECD−NEA demand used in the earlier example is shown in Figure 10−4. The cumulative uranium consumption is shown as a function of time for the various degrees of thorium utilization in a breeder that in turn provides the fueling requirements for denatured thorium cycle LWRs. Because a fixed demand must be met, and the capacities of both reactor types are constrained by fissile supply—plutonium in the breeder case and uranium-233 for the LWR—additional uranium-235 fueled LWRs are required with most symbiotic combinations. Where these are

Figure 10−4. Cumulative Uranium Consumption as a Function of Time for Various FBR Alternative Cycles *(Symbiosis with Denatured LWR)*

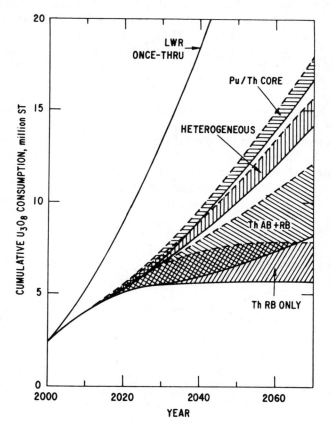

required, U_3O_8 consumption rises. Cases are shown extending from thorium usage in the breeder radial blanket only, through axial blankets, internal blankets in a heterogeneous design, and finally, in totally displacing uranium–238 as the fertile material. For each geometric variation, the range corresponding to fuel type—oxide, carbide and metal—is indicated by shaded areas.

There are overlaps between fuel types, but, in general, the uranium requirements are dictated most strongly by the degree of the thorium utilization in the breeder. The uranium requirements increase progressively in the amounts shown in Figure 10–4, as more thorium is incorporated into the fast breeders, from thorium in the radial blanket only, to the full thorium fertile material core.

The ratio of breeders to dispersed converters is the other important factor. Normal economic and engineering pressures will tend to maximize the number of converters in such a symbiotic system because of the freedom in deployment that, in this scenario, only the converters enjoy. However, specifying an increase in the ratio of dispersed converters to secured breeders translates directly into an increase in U_3O_8 requirements, if the demand is to continue to be met.

The maximum converter/breeder ratio allowed by each of the variants in Figure 10–4 is shown in Figure 10–5; again plotted as a func-

Figure 10–5. The Ratio of Dispersed Converters to FBRs as a Function of Time for Various Alternative Cycles *(Symbiosis with denatured LWR)*

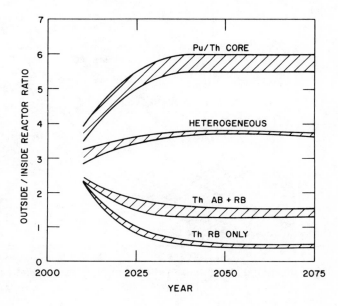

tion of time. The converter/breeder ratio is termed the outside/inside ratio in the figure, in line with the secure area notion. The outside converters include both isotopically denatured thorium cycle LWRs and those standard uranium cycle LWRs, fueled from additional mining of U_3O_8 that are necessary to meet the assumed demand. The shaded areas again show the effect of breeder fuel types.

The trade-off between reactor ratio and uranium consumption is shown explicitly in Figure 10−6. Here the asymptotic reactor ratios from Figure 10−5 are plotted as a function of their corresponding U_3O_8 consumption of reactor ratios and uranium consumption possible with design variations, including fuel types, for each of the breeder reactor thorium configurations. The advanced fuel types show significant uranium improvements for both the thorium external blanket cases, with the improvement lessening as thorium is introduced in the core.

For any given reactor configuration and fuel type, the trade-off between reactor ratio and uranium consumption is direct. There is no quantifiable optimum. The balance depends on the relative weight given to the two factors. In the secure area context, it depends on

Figure 10−6. Relation between the Outside/Inside Reactor Ratio and the Uranium Requirements for Various Fuel Types and Fuel Cycle Options

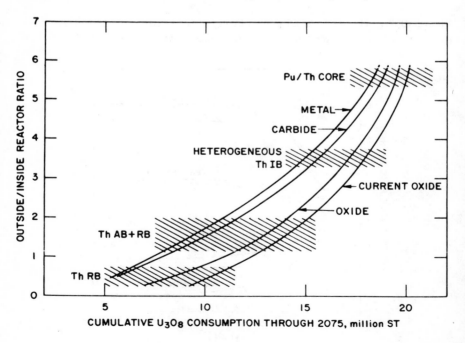

the relative importance assigned to confining breeders to such areas. Qualitatively only, the reactor ratios should be maximized without inducing unacceptable penalties in uranium resource requirements. For the reference demand Figure 10−6 quantified the possibilities.

SUMMARY

1. Present-day reactors use uranium inefficiently; this gives rise to concern about the adequacy of uranium resources to meet expected nuclear demand for very long. Really substantial increases in efficiency of uranium utilization require reprocessing. Reprocessing activities give rise to concern about their possible use in fission weapons acquisition.

2. The basic properties of nuclides severely limit both the number of alternative ways that fuel utilization can be improved and the amount of the improvement that is possible from any of the alternatives. By far the greatest improvement comes from plutonium use in a fast reactor. The properties that allow this are peculiar to plutonium.

3. There are basically only two fuel cycles that can be considered as alternatives to the plutonium–238/uranium fuel cycle. One is a uranium–233/thorium fuel cycle, a cycle that is very similar in requirements, including reprocessing, to the plutonium–238/uranium cycle. The other is continuation and refinement of the current once-through cycle. Basic nuclear properties probably rule out the use either of thorium or of a fast spectrum in a once-through cycle. As these two are the only bases for substantial fuel utilization improvements (the first in thermal reactors and the second in fast reactors), the improvements possible in once-through are severely limited.

4. A small number of technical measures to increase proliferation-resistance have been proposed. So far, they have concentrated largely on differences in fuel form. The degree of resistance they offer in contrast to the degree of difficulty they introduce in producing power is the central issue in the consideration of these possibilities.

5. Improvements of an institutional nature are of two types. The first are improvements in international safeguards—most importantly, nuclear materials accountancy—essentially strengthening or augmenting current IAEA procedures. The second involves agreements between nations to limit distribution of sensitive technologies and to multinationalize or internationalize sensitive elements of the fuel cycle.

6. One example of a scenario that includes both such technical measures and institutional arrangements to effect a balance between energy generation and proliferation concerns is the secure area/ symbiotic cycle suggestion. It takes advantage of the superior fuel properties of plutonium while limiting the distribution of fissile material to forms that may have improved proliferation resistance.

REFERENCES

OECD Nuclear Energy Agency. 1978. "Nuclear Fuel Cycle Requirements and Supply Considerations, through the Long-Term." February.
_____ . 1977. "Uranium Resources, Production and Demand."

Chapter 11

Nonproliferation Criteria for Assessing Civilian Nuclear Technologies

Henry S. Rowen

Two trends are affecting the spread of nuclear weapons. One is the growing access to readily fissionable materials as a by-product of the spread of civilian nuclear technology. This access is not inhibited by existing international agreements, which do not preclude states that do not possess nuclear weapons from having uranium–235, uranium–233, or plutonium in a form immediately usable in explosives.

The second is the fact that many countries acquiring easier access also have an increased incentive to acquire nuclear explosives, or at least to shorten the lead time to them. These are countries that are often involved in conflicts or tensions with neighbors; many with an increased capacity to make nuclear weapons are outside of protective alliance systems or fear weakening of their alliance ties.

These trends, which both increase the demand for nuclear explosives and lower the difficulty of obtaining them, do not lead inescapably to a prediction that all the nations capable of taking the additional steps from civilian uses of nuclear energy to nuclear explosives—around forty by 1990—will do so. There are inhibiting factors at work in many countries. But the existence of this "overhang" of countries that can be legally (that is, without violating international agreements) very close to possessing nuclear explosives makes urgent a reassessment of the structure of international nuclear cooperation. Such a reassessment is being undertaken in the International Fuel Cycle Evaluation program. The U.S. Congress has just passed legislation that reflects a reassessment of its own, requiring as its most important provision that recipients of nuclear technology and materials from the United States pledge not to develop nuclear explosives using materials from any source. Important as these changes are, they still do not preclude countries with whom the United States conducts civilian nuclear trade from being close to having nuclear explosives, without violating any agreement. They merely provide that nuclear materials from the United States cannot be "reprocessed or otherwise altered" without prior approval of the U.S. government. This leaves other paths open, including imports from other suppliers who may impose less stringent conditions or indigenous development not involving use of materials from the United States.

If nuclear energy is to continue to be developed and find wider use throughout the world without bringing in its train heightened and possibly intolerable risks, there is an essential task to be performed: it is to design mixes of nuclear technologies and political arrangements directed towards three types of objectives.

1. Strengthening incentives in states lacking nuclear weapons for not acquiring nuclear explosives or drastically shortening the lead time to make them.
2. Promoting international agreements and choices of technology that will leave states some distance from having nuclear explosives; that is, producing situations likely to be more stable in the face of crises and conflicts than is likely with the current international rules and technologies.
3. Producing economically competitive supplies of energy and also helping to assure supplies of energy to countries conforming to international agreements limiting access to nuclear explosives.

For a nonproliferation strategy to have much chance of working, it must take into account the great diversity among nations, which includes large differences in the following aspects:

1. Size, economic and technical sophistication, and economic growth rates,
2. Perceived threats to security; also differences in the support provided by international ties, such as the European Economic Community, NATO, and bilateral alliances,
3. Dependency of several kinds, including dependency for nuclear supplies or for other fuels, dependency derived from trade ties or economic aid, and military security dependencies. Governments also vary in the importance they attach to reducing dependencies, for instance on supplies of energy from abroad.

Despite this diversity, most states are alike in their sensitivity to nuclear policies advanced by the weapon-possessing states that appear discriminatory. This was evident in the states without nuclear weapons during the Non-Proliferation Treaty (NPT) negotiations. They objected to the possibility that they might be commercially disadvantaged and therefore insisted on the inclusion of Article IV. There will be reluctance, and possibly strong opposition, to the adoption of nuclear systems and rules that are not broadly or perhaps universally applicable. However, one distinction is unavoidable in a nonproliferation strategy—that between weapon-possessing states and those without nuclear weapons.

THE MAIN INFLUENCES

On the demand side, the main influences on opting for nuclear weapons include perceived threats from regional adversaries and the consequent increase in the demand for military strength and perhaps for nuclear explosives. This incentive is heightened for countries that are isolated internationally or that perceive their security ties weakening. It is also heightened if their adversaries have nuclear explosives. A push to get weapons might be felt strongly even if a regional adversary is not estimated as having bombs now, but is seen able to acquire them suddenly. Fear of sudden moves toward nuclear weapons creates a kind of preemptive instability familiar in analyses of circumstances that might prompt the *use* of nuclear weapons.

This suggests the existence and, in time, the realization of the underlying common interests that neighboring states have in measures that would increase the time required to get bombs.[a] It is a mistake to hold that all of the factors working on the demand side are for the acquisition of nuclear explosives. Alliance protection, opposition by allies, domestic opposition, lack of perceived threats, and fear of stimulating others to go ahead are among the reasons that most of the nations capable today of having nuclear weapons do not possess them. Instruments of influence that might be used to discourage such acquisition include limiting or ending conventional arms transfers, economic aid, investment, and trade or security guarantees. There may also be the worry that nuclear weapons might become a factor in internal disturbances and even that they might escape governmental control. And finally, there is concern that any future conflict would become enormously destructive.

[a]This formulation is similar to nuclear-free zone agreements, for instance in the treaty of Tlateloco, but goes further by proposing prohibitions on immediate access to nuclear explosive *materials*. Such access is not prohibited under the NPT, Tlateloco, or other current nuclear-free zone proposals.

An examination of the particular situations of a number of third world countries suggests that great power agreements on limiting nuclear systems (e.g. SALT reduction, CTBT) appear to be largely irrelevant in the foreseeable future. In some cases, agreement between the great powers could have an adverse effect (that is, in Europe, especially with respect to some SALT agreements).

The principal influences on the *supply* side include the *incremental* costs of steps to acquire nuclear explosives—that is, costs in addition to those incurred in programs undertaken for nonmilitary purposes. It may be useful to think of three classes of nuclear-related activities:

1. Those that are a necessary part of a civilian nuclear program. There are many joint products (those useful for military as well as nonmilitary programs) from purely civilian activities. Military benefits come free up to a point.
2. A gray area in which actions of small, doubtful (perhaps negative) civilian economic benefit are taken that also advance nuclear weapons capabilities.
3. Unambiguously military steps that cannot be associated with any plausible civilian use.

The present rules (bilateral agreements for cooperation and the NPT) offer civilian "covers" for the further incremental steps that might be undertaken for military purposes.

The existence of "covers" of varying degrees of plausibility makes the interpretation of signals of possible military programs difficult to interpret. It also makes it possible for a government moving towards bombs to have to undertake fewer activities covertly and to introduce covert activities at a later stage of development.

The possible variation in civilian starting points affects the incremental capital and labor needs for military purposes. It is also affected by the size and sophistication of the military programs for such system components as delivery systems, protection of forces, surveillance, command and control, and the like.

Nonmonetary costs are also incurred in the form of risks associated with having or moving towards nuclear explosives. These risks will be affected by anticipated reactions of neighbors and others as discussed previously. Because possession of nuclear explosive materials is legitimate, such possession can be represented to neighbors as a "civilian activity." To the extent that this position is accepted, it lowers the risks of being close to having bombs. For example, the present rules accept as legitimate the possession of research facilities that use highly enriched uranium or produce sizable quantities of plutonium, independently of the expected research output from these reactors.[b] The recycling of plutonium in LWRs has been argued for, on the grounds of commercial viability or fuel independence or getting ready for the imminent introduction of the breeder, despite the weakness of these arguments on economic grounds. In short, some economic "covers," even if threadbare, can be found for almost any nuclear activity of military utility.

Because of variations in costs of fossil fuels, costs of capital, size of electricity grids, and concerns for security of supplies, economic tests are often not easily made. However, there is no way to escape applying an economic test in formulating new rules for the use of nuclear technology. Activities in the third of the preceding categories, those of an unambiguous military character, need to be outlawed if there is to be a nonproliferation regime at all. And so do dangerous activities in the second category, where there is little or no economic benefit, or where the economic benefits are small in relation to the dangers. Fortunately, it appears that none of the most dangerous activities at the present stage of nuclear development are in the first category—those clearly essential to the use of nuclear energy.

The civilian and military joint products and gray area ambiguities bear on the concept of "critical time," the time required to move from an agreed, civilian safeguarded status to one in which a given number of nuclear weapons, or a rate of production of them, is achieved. The more the civilian-military distinction is blurred (the better the civilian cover), the shorter the critical time can be made.

IMPLICATIONS FOR NONPROLIFERATION STRATEGIES

Nonproliferation strategies might seek to influence the demand for nuclear explosives through improved alliance ties, regional security associations, and nuclear free zones, as well as the ease of access to explosives through agreement on increasing the difficulty of each access through changes in international agreements on technologies, or through a mix of such measures. The discussion that follows focuses on a *supply-oriented strategy*, not because such a strategy by itself is likely to be optimal, but because it would be a significant component of a broad strategy, and it is the one that has been cen-

---

[b] Efforts are being undertaken to reduce or eliminate the presence of highly enriched uranium in these reactors.

tral to the nonproliferation efforts of the United States in the past several years.

A supply-oriented strategy could have two components:

1. A set of incentives for choosing less dangerous nuclear systems instead of more dangerous ones (and in some cases the choosing of non-nuclear rather than nuclear technologies);
2. A set of political agreements restricting especially dangerous systems or components of systems.

For such a strategy to have a prospect of being effective, it should encompass all of the paths to a bomb from a legitimate safeguarded state. Specifically, it should include:

1. Paths starting from large plutonium reactors, including those labeled research reactors.
2. Isotope separation technologies.
3. Power-reactors-related paths, based on using either
    a. Material available at the front end, or
    b. Material available at the back end.
4. Various possible future technologies, such as accelerator breeders or fusion-fission technology.

In order to assess the properties of different nuclear systems, we need a framework for analysis such as the one presented in the next section.

THE KEY PROLIFERATION-RELATED ATTRIBUTES OF NUCLEAR SYSTEMS

The principal parameters in a simple model are these:

*   Description of a civilian nuclear technology status that is compatible with international undertakings.
*   Description of a military status—the possession of, say, three or fifty fission bombs with yields in the 1–20 KT range, or a given production rate of such weapons.
*   The time required to take the physical actions to move from civilian to military status—"critical time."
*   The probability that one or more important other governments will detect signals of the move, and interpret them, and be able to act before military status is achieved.
*   The warning time available to other governments before military status is reached.

- The incremental costs of the constraints imposed by the international agreements on the civilian status as compared with an unconstrained system.
- The incremental costs of moving from civilian to military status, including program costs, the conditional consequences of failure to achieve military status within the critical time, and the conditional consequences of giving warning to outsiders.

In the case where warning time is less than the critical time, if the latter is small enough, the warning would be effectively zero. Furthermore, even in the case where the critical time is fairly long, detection does not necessarily mean generating usable warning, given the problem of ambiguity in interpretations. Also, the smaller the warning time, the fewer the responses available to the world community, and hence the smaller the estimated cost of moving from civilian to military status to the government in question.

Probability of success will be higher if the steps from civilian to military status are few and simple and if the critical time is short because fewer (illegal) steps have to be taken, since each step contains the possibility of mistakes and delay. Success probability also varies greatly with the quality of the human and material inputs available for the move.

The expected costs of moving to military status, in addition to resource costs, include those associated with being caught before getting there and program failure (at least failure to achieve military status in a specified time, such as the critical time). The relative weight of these costs is, of course, very situation-dependent. In an intense conflict, much greater weight is likely to be given to the costs of failure to achieve military status than in being observed moving to that state; in other situations the opposite may be true. A risk-averse government might judge the costs to be too high to warrant making the move. On the other hand, a less risk-averse government or one that perceives high costs in *not* making preparations for getting the bomb may proceed even if the expected cost (probability of giving usable warning, times the costs of responses by others) is high. Possible consequences of trying to get the bomb and failing include failing to deter an adversary's attack or triggering a more violent action than otherwise would have occurred.

Warning could come via national means of verification or via IAEA channels. The lag times and, indeed, the reliability of transmission of signals in the latter channel make it unreliable for small critical times, but perhaps useful for long ones. Warning is also a function of the "noise" created by civilian activities. For instance, it is likely to be a good deal easier to conceal a small reprocessing "laboratory" for separating out pure plutonium, if sizable activities involving large numbers of trained people are legitimately engaged in a similar activity, such as coprocessing. The probability of warning of covert activities can be thought of as consisting of two components, one independent of the size and character of the program involved in moving to military status, and the other dependent. Warning from the former source might come via human intelligence, for example; from the latter it might come via technical means of information gathering.

There are trade-offs among these parameters. A government contemplating the move might try to minimize critical time within some probability of success constraint. Alternatively, the objective of a low warning probability within some critical time and success probability constraints might be sought. An objective of a nonproliferation strategy is to try to shift these parameters into values that will cause governments to prefer to remain in civilian status.

Governments might seek an intermediate state between civilian and military status in which they do not have immediately usable nuclear explosives, that is, to shorten the lead time. This would, nonetheless, be a violation of agreements.

Critical time and success probability are very dependent on the state of technical development of the country and, especially, on the skill level of the technical people doing the work. For instance, if reprocessing activities are excluded from the civilian status, the time required to build such a plant (intended to be secret) and process material would depend heavily on whether the operators had had relevant experience elsewhere.

SOME ILLUSTRATIVE CASES[c]

Case 1.   Assume that the civilian status corresponds to today's system. States without nuclear weapons can have separated plutonium of high enriched uranium (HEU) in metallic form and facilities for shaping it for criticality experiments or fabricating highly concentrated fissile fuel. Associated with this status are the following key parameter values, assuming military status is defined as the acquisition of five to ten assembled 1–20 KT yield weapons. Assume, also, that those non-nuclear components with long lead times can be "covered" by the other military—or even civilian—programs and have been completed at the time the move occurs.[d]

[c]Taken from Wohlstetter et al. 1979.

[d]This raises the question as to whether the civilian status excludes work relating to the non-nuclear components of nuclear explosives. Such a prohibition

Critical time:	nearly zero
Success probability:	high (depends only on the skill of the bomb designers and the adequacy of the non-nuclear components)
Warning probability:	low, likely to be zero
Warning time:	low, likely to be zero
Costs of the civilian constraints:	zero
Incremental costs of the move:	nearly zero in resources; also in costs of response by outsiders before reading military status (given that warning time is nearly zero)

Case 2. Civilian status is limited to plutonium and highly enriched uranium in nonmetallic forms, such as plutonium nitrate in dioxide or spent fuel; or uranium in the form of highly enriched uranium hexafluoride. This status permits reprocessing plants and enrichment plants of any type.

Critical time:	days to weeks to rework material to make it weapon-ready; however, six months to a year are needed for preparation of rework facilities and nuclear components; if such facilities are a violation, then critical time is months to a year
Success probability:	high, if operators are skilled; low (or longer critical time), if less skilled operators available
Warning probability	very low, if rework facilities legal; probability still low even if such facilities prohibited
Warning time:	low if rework facilities permitted; higher—but probability still low—if rework facilities are prohibited
Costs of civilian constraints:	~zero
Incremental costs of the move:	resource costs, 1-10 million dollars for reworking plutonium or uranium hexafluoride; with warning times low, perhaps near zero; the cost of expected responses of outsiders would also be low

---

seems appropriate for inclusion in revised agreements, although there are serious questions about the verifiability of such prohibition. The assumption in Case 1 is that non-nuclear work has been done when the move from civilian to military status begins.

Case 3.   Civilian status limited to materials no nearer to use as explosives than mixed oxide fuel (MOX) or low enriched uranium (LEU). These materials might be shipped from outside or produced in own reprocessing plant designed for coprocessing spent fuel. This status would exclude isotope separation technologies for quickly producing HEU from LEU or a plant for producing a separate plutonium stream.

Critical time:	several months to a year for the construction of a fresh fuel reprocessing "laboratory" (with extra time to rework the material, of days or weeks) (However, the possibility might exist of changing the coprocessing plant flows to provide a pure plutonium output.)
Success probability:	high for a skilled team; moderate for an unskilled one
Warning probability:	moderately low, based on construction of reprocessing "laboratory"; perhaps high if based on diversion of MOX
Cost of civilian construction:	~zero
Incremental costs of the move:	resource cost, $1-$10 million (including cost of facilities for separating plutonium from fresh MOX fuel); low warning probability suggests low incremental costs in other categories

Case 3a.   In a variant of Case 3, civilian status might be limited to materials that are harder to handle than unirradiated MOX. In this case, only irradiated or spiked MOX would be permitted (as in the CIVEX proposal of Marshall and Starr). In this case, critical time is higher as is the incremental cost of moving to military status. So also is the cost of the civilian status.

Case 4.   Civilian status is defined as LWRs, AGRs, CANDUs, or research reactors using slightly enriched or natural uranium fuel only once, with safeguarded spent fuel storage.

Critical time:	18–20 months for construction of a facility to extract plutonium from hot spent fuel plus months to a year to process material (or weeks to months if skilled personnel are available)

Success probability:	for a processing period as short as a few weeks, the probability might fall well short of unity even with skilled people; within several months a skilled team would achieve a success probability near unity, but not an unskilled team
Warning probability:	moderate to high
Warning time:	given the larger size and longer times involved, warning time is likely to be significant, perhaps on the order of weeks to months
Costs of civilian constraints:	at today's uranium prices and reprocessing costs, zero
Incremental costs of the move:	resource costs, $10–$100 million; also possible high costs from response by other nations

Case 5. Civilian status as in Case 4, but the path from civilian to military status chosen is the construction of large reactors to make plutonium together with separation facilities. (Large plutonium-producing research reactors are assumed not permitted in the civilian status.)

Critical time:	~5 years
Success probability:	high; a long time to train people is available
Warning probability:	high
Warning time:	also high; these facilities are larger and harder to conceal
Costs of civilian constraints:	the incremental costs in this case are not related to civilian purposes
Incremental costs of the move:	resource costs, $10–$100 million; a high possible response cost from others

Case 6. Civilian status as in Case 4, but the path to military status chosen is construction of a plant to produce HEU. With present technologies:

Critical time:	$\geq$ 5 years
Success probability:	high for technically advanced countries, but low for others
Warning probability:	high
Warning time:	high; these are generally large facilities with current technologies (With future technologies, warning time may be lower.)

Costs of civilian constraints:	the exclusion of enrichment plants capable of quick conversion to HEU production might be regarded as imposing a cost by increasing energy supply vulnerability (On the other hand, less costly ways of providing supply assurances other than building small national enrichment plants seem available.)
Incremental costs of the move:	resource costs assuming starting *de novo* (not converting existing LEU producing plan): $100 million to $1 billion.

FURTHER EXAMINATION OF THE "BENCHMARK" CASES OF CURRENT REACTOR TYPES OPERATED ONCE-THROUGH

The "benchmark" corresponds to Cases 4, 5, 6, and possibly 3a above. This standard would, by agreement, make it a violation to receive plutonium or HEU (above some threshold amounts for basic research) in fresh fuel or to have reprocessing facilities or enrichment facilities. It should also provide for the removal of spent fuel after a short cooling period.

The construction of a reprocessing plant would be a violation with a potential critical time—and warning time—of a year or more. (If the construction of the plant were judged not to have a high probability of giving warning, the critical time—and possibly warning time—would be several weeks.) This is distinctly superior to prospects with the lower numbered cases.

The cost to civilian nuclear applications is low, given the current cost estimates for reprocessing versus fresh uranium fuel. So long as enriched fuel is reliably available, there is no significant economic penalty from an enrichment technology restriction. (However, there may be concerns about security of supply and there would be incremental costs from offering fuel assurances.)

This benchmark case excludes the wide distribution of isotope separation facilities and large plutonium-producing research reactors. (Research reactors have been widely distributed although the large ones, so far, are in relatively stable states.) Until recently, enrichment activities were confined to states with nuclear weapons, but now they are being carried out, at least on a small scale, in the Netherlands, Japan, and South Africa; Brazil is scheduled to join this list. Moreover, a desire to increase confidence in the availability of enrichment services argues in favor of having a number of politically independent suppliers. One way of reconciling these partially conflicting objectives is to make such plants, at least in nonweapon states, subject to multinational ownership and operation, and to use technologies with a long equilibrium time from LEU to HEU production.

Removal of spent fuel to safeguarded storage sites seems necessary in Cases 4 and higher. Also, storage under multinational auspices, such as IAEA, should be considered.

The benchmark might be approximated in Case 3a, in which a radiation barrier is incorporated in the fresh fuel. To reflect preferences of most governments for security of fuel supply, the radiation barrier would have to be designed to allow for several years' storage of fresh fuel before use.

THE ANALYSES APPLIED TO FUTURE TECHNOLOGIES

The benchmark cases are readily adapted to the concept of energy centers, a small set of places where activities would be carried out. Within energy centers, spent fuel might be stored, large plutonium research reactors could be located, or isotope separation facilities (at least those with a short critical time for HEU production) be located. The concept of such centers requires making a geographic distinction in the distribution of nuclear activities. Nonproliferation objectives, in general, seem to require that the handling of weapons materials be done only in weapon-possessing states. However, there are gradations in the proliferation risk associated with various fuel cycle components, and some, such as enrichment technologies with a long critical time, might be suitable for energy centers not in states that have nuclear weapons.

In any case, the issue of discrimination between states permitted to have, or be, energy centers, and those not, is likely to be a large obstacle to the realization of some of these variants. It was, after all, concern about perceived discrimination that caused Presidents Ford and Carter to decide to defer commitment to plutonium recycle in the United States. It is by no means clear that many, if any, states are willing to have their territory effectively designated as "remote" from an energy standpoint. At the least, mechanisms would have to be devised for assuring nuclear fuel, and sharing equitably any economic advantages from being a center.

No significantly different reactor technology from that now in use is likely to be introduced until after the year 2000 at the earliest. Plutonium breeder programs are proceeding slowly; their economics are uncertain and may turn out to be unfavorable. And although the use of the thorium/uranium–233 cycle could begin somewhat earlier, a good deal more work evidently needs to be done on uranium-233 reprocessing.

Case 7. Use of uranium-233 denatured with uranium-238 or LEU (as in Case 4), outside of energy centers, also enrichment technology with short equilibrium times or reprocessing plants outside of the energy centers.

Critical time:	probably months to construct a facility to extract the plutonium (present in low density) from hot spent fuel, or to build isotope separation facilities to separate uranium-233 from uranium-238
Success probability:	hard to estimate for a distant time period. (However, even then probably not all governments will have access to highly skilled teams.)
Warning probability:	high if substantial facilities involved; low if, for example, laser isotope equipment is widely available for legitimate purposes
Warning time:	see comments on warning probability
Cost of civilian constraints:	unclear; among other things, the cost of the thorium cycle is uncertain
Incremental costs of the move:	uncertain; perhaps $10–$100 million

Case 8. LMFBRs widely distributed with use of coprocessing and shipment of MOX. This case corresponds to Case 3 above (or Case 3a, if only irradiated fuels are made).

CONCLUDING OBSERVATIONS

The present rules permitting access to metallic plutonium and HEU promise little warning of acquisition of, at least, a small stockpile of weapons. With current technology, there is a possibility of being able to increase critical time substantially, so that anticipated warning times would allow for preventative actions. But this is not the same thing as having assured warning time of comparable length. The cost to civilian activities of accepting such constraints is low for most of the cases, and all of them currently applicable. Objections on grounds of energy independence need to be met through design of suitable system modifications. The principal changes in current practices needed seem to be in fuel assurance, in disposition of spent fuel, and in revision of agreements for cooperation and in safeguards.

In the future there could be a widespread shortening of critical times from the spread of enrichment technologies. If widespread distribution of MOX is permitted for use in thermal or in fast reactors, critical times may well be too short, that is, less than the warning time. Schemes for extending uranium resources, by adopting some variant of the thorium cycle (operated on a denatured cycle) or through the plutonium breeder using radioactive fresh fuel, have the potential for extending critical times, but it is not yet clear by how much, nor is it clear what the costs of such systems would be.

REFERENCE

Wohlstetter, Albert, et al. 1979. *Swords from Plowshares.* Chicago: University of Chicago Press.

❋   *Chapter 12*

The International Dimension
of Nonproliferation Criteria

Lincoln Gordon

The foregoing papers by Drs. Till and Rowen are so rich in
analytical ideas and policy implications that a comment
limited to fifteen minutes can scarcely do them justice.[a]
My observations will touch briefly on a few specific points in their
papers and will outline the major aspects of the international dimension in decisionmaking on nonproliferation criteria.

Dr. Till has devised some highly ingenious presentations in two
dimensions of a set of multidimensional interrelations, including resource base estimates, nuclear energy growth assumptions, resource
consumption for alternative fuel cycles, and related lead times. What
he omits (not from negligence, but from the impossibility of putting
everything into two dimensions) is the economic factor: the relative
costs of the alternative fuel cycles and their relation to assumed
prices of uranium and thorium. A minerals economist is bound to
boggle at the two fixed resource base assumptions (2.4 and 4.3 million tons of U_3O_8, as if mineral depletion were like emptying a glass
of water). There is always *some* expansion of the resource base at
higher costs and prices, although there can be much argument among
geologists whether that is a smooth function or one subject to sharp
discontinuity because of different kinds of geological occurrences.

It must also be borne in mind that the Till analysis refers primarily
to the United States. On the resource side, it is a question of great
importance whether our planning should be based on the hypothesis

---

[a] This discussion refers both to Dr. Till's contribution to this volume (chapter
10) and to a published paper which was circulated to participants before the
Argonne Symposium (Till and Chang, 1978).

of zero imports and exports. And with respect to nonproliferation criteria and the time-phasing of alternative fuel cycles, especially the deployment of some kind of breeder, the critical question is how this kind of analysis would appear globally or in the eyes of other countries or regions anticipating substantial reliance on nuclear power during the first half of the coming century.

The policy implications of Dr. Till's paper are of great interest in three respects: (1) the distinction between sensitive activities permitted in "protected fuel centers" and less sensitive activities elsewhere; (2) the limitations on proliferation resistance of thorium cycles resulting from the greater ease of uranium–233 isotopic separation compared with uranium–235 separation; and (3) the urgency of developing plans for some sort of advance beyond the once-through light water reactor if nuclear energy is to be a major factor after the first quarter of the next century.

Dr. Rowen's paper deals with a wider array of issues, touching on motivations for weapons proliferation as well as means and costs, and relating the proposed nonproliferation "benchmark" to the character and location of future nuclear technologies. I will note specific points of agreement and disagreement with it as I summarize my own view of the international dimension.

UNCERTAINTY AND THE END OF UNILATERALISM

Nuclear energy policy is an extreme example of the problems of decisionmaking in the face of uncertainty, even more so than most other energy policies because of the enormous lead times, the huge size of the required investments, and the special constraints of proliferation dangers. Within the next few years, decisions will have to be made for or against power technologies that will come into large-scale use only after 2025, based on some estimate of probabilities of requirements and costs. Yet even for the period from now until 1985 or 1990, the spate of conflicting energy projections demonstrates uncertainty on every relevant parameter: rates of economic growth; overall energy demands; electric power demands; the nuclear segment of electric power; availability and prices of oil, coal, and uranium; relative environmental acceptabilities; and the relevant costs and proliferation risks of alternative nuclear fuel cycles. As the time span lengthens, further uncertainties are introduced by the possibility of new technologies, such as direct and indirect solar energy, again with presently uncertain capital and maintenance costs and environmental impacts.

Were it not for the proliferation hazard, prudent planners would almost certainly include nuclear breeder development at least to the point of potential commercialization, even if only as a protective hedge in this welter of uncertainties. In face of a continuing demand for some expansion of electric power (even at rates far below those of the past), for replacement of obsolete units, and for gradual phasing out of oil and gas as boiler fuels, the penalty for underbuilding capacity would be much larger than the costs of overbuilding. Notwithstanding the debates on the merits of economic growth in general, and nuclear energy in particular, it seems to me plausible that if plutonium were not a weapons material, Barnwell would be proceeding at present and so would some kind of demonstration breeder—although my expert friends persuade me that it could well be a better design than Clinch River. Whether and when recycling would take place in converter reactors under those circumstances would be a purely economic decision, depending on actual costs of reprocessing and prices of uranium and enrichment services.

So the proliferation issue is critical. But it should be kept constantly in mind that meaningful nonproliferation policy can no longer be made by the United States alone. Some discussions of public policy in this field since 1974, including congressional proposals for export control, seem like a throwback to 1946 and the McMahon Act. Today or soon, a country intent on nuclear energy development will be able to look to multiple sources for reactors, enrichment services, technological training, reprocessing offered on toll, and perhaps even waste management.

For some purposes, therefore, agreement among all potential suppliers is a minimal condition of effective policy, and for other purposes there must be agreement among both suppliers and recipients.

ASPECTS OF THE INTERNATIONAL DIMENSION

In each of the following six respects, my assessment of the international dimension differs to some extent from that given in Dr. Rowen's paper. Some of the issues will be presented in question form to invite discussion.

1. Economic and Security Trade-offs in the Eyes of Others

It is commonplace that industrial countries with large energy requirements and limited domestic resources of both fossil fuels and uranium will appraise the nuclear breeder differently than will the

United States, Canada, Norway, or Australia. If countries like Germany or Japan were applying Charles Till's method to their situations, what figures should they use for a dependable uranium resource base? They might agree that breeders provide no economic advantage at today's uranium and reprocessing costs (see Dr. Rowen's Case 4), but the relevant comparison is expected costs in the decades after 2010. Given the current experience with OPEC and the uncertainties of Australian, Canadian, and American uranium export policies in recent years, is it not entirely reasonable for them to place a high premium on greater energy resource independence than Case 4 would permit? What international institutional arrangements might respond effectively to the desire for energy resource security? For the more advanced developing countries like Korea and Brazil, sometimes referred to as the New Japans, may not similar attitudes legitimately prevail apart from any possible interest in weapons?

2. Motivation and Means for Acquiring Nuclear Weapons

Dr. Rowen correctly analyzes the interweaving of motivation and means, and calls for political action to restrain proliferation on the demand side, as well as technological action on the supply side. But his paper opens with the puzzling assertion that many countries have an increased incentive to acquire nuclear explosives. That seems to me questionable. External incentives depend on changing regional power situations and perceived threats, while internal ones may reflect the character of regimes. Compared with a decade ago, I would argue that the incentives today are greater in Korea, South Africa, and Iran, but less in Pakistan and Spain, and unchanged at nearly zero in Latin America.

At the same time, I would amplify Dr. Rowen's point about easier access because of the multiplication of sources of nuclear equipment and technology, and also the likely trend of further technological development, especially in isotope separation. Even for chemical reprocessing and remote handling techniques, there are more trained people in the world, including those disemployed at abandoned plants in Europe and North America, who might be offered employment in other countries.

3. The Limits of Discrimination

In the politics of international relations, there is nothing harder for a nation to swallow than overdiscrimination, placing it in formal second-class status. The Non-Proliferation Treaty itself is a remarkable exception, since its very essence is the distinction between states with nuclear weapons and those without. But it also entails a global bargain: in return for foreswearing weapons acquisition, states without weapons are supposed to enjoy full access to technology for peaceful purposes. Where peaceful and military purposes intersect, as with uranium enrichment and plutonium separation, the legalities are inherently ambiguous. The politics, however, require that discrimination in nuclear power arrangements be reduced to an absolute minimum; otherwise, the threshold nonweapon states could not be kept within the system. Dr. Rowen's paper (p. 191) attributes the resistance to discrimination to fears of commercial disadvantage. That is one aspect, but the roots go much deeper, including the desire to participate in modern technology and even profound psychological concerns with human dignity.

That issue rubs on two specific points: (1) whether international safeguards and an international rule against national control of sensitive facilities (enrichment and reprocessing) should apply to the civilian activities of states having weapons as well as states without, and (2) whether the location of multinational fuel centers should be limited to states with weapons. It also affects the extent and character of participation in multinational centers.

4. Interrelation of Technological, Political, and Institutional Factors

That brings me to the interrelations among technological, political and institutional elements in nonproliferation criteria. Dr. Rowen's paper (p. 184) refers to incentives for choosing less dangerous nuclear systems, or non-nuclear alternatives, and to political agreements restricting dangerous components. His proposed benchmark limits "civilian status" to once-through converter-reactor cycles. For the immediate future, that position is very persuasive. But an international nonproliferation system design has to anticipate probable changes in technology and the probable course of technological diffusion and to take into account a variety of different national appraisals of trade-offs involving energy security and long-term costs of alternative energy supply systems.

The concept of multinational energy centers could reduce the element of formal discrimination and also reduce the inducement to weapon status arising from the fear that a potentially hostile neighbor will get there first. On the negative side is the possibility of accelerated technological diffusion through the multinational staffing of the centers. In my judgment, the advantages of establishing a multinational system very soon, before sensitive facilities under national control in several states without weapons become both irreversible *faits accomplis* and precedents for other states, outweigh the risks of marginal acceleration in technological diffusion.

I do not believe, however, that their location could be confined exclusively to weapon states, as Dr. Rowen's paper appears to suggest (p. 191). The danger of host country takeover could be minimized by selection of locations in small countries, not conceivable candidates for great power status and vulnerable to sanctions from coparticipants, and also by suitable technical arrangements for active use denial.

5. The Force of the American Example

Dr. Rowen's paper attributes the Ford and Carter Administrations' decisions to defer plutonium reprocessing within the United States to "concern about perceived discrimination" (p. 191). I would rather describe it as reflecting a hope that the force of the American example would persuade others to revise their attitudes, or at least a conviction that efforts to persuade others to cease and desist could not be effective if we were proceeding ourselves. In my view, this decision was correct and useful for a limited period of time, but its usefulness will diminish rapidly. Its failure to affect plans in Europe and Japan reflects the differences in national situations and assessments already mentioned.

6. The Need for Collective International Decision

From this summary review of the international dimension, it follows that the title of our symposium is inadequate for nonproliferation policy. Nonproliferation is not merely a national energy issue on which "we" are able to decide. It is an international issue on which the deciding "we" must become the entire relevant world community, namely all governments with a significant present or future interest in nuclear energy.

In historical perspective, it is noteworthy how the scope of nonproliferation policy discussions has steadily broadened over the years. In the 1950s and 1960s, it shifted from unilateral American action to joint American–Soviet initiatives, of which the most important was the Non-Proliferation Treaty. In the early stage of post–1974 reassessments, the Suppliers' Group was called into being. It was followed a year ago by creation of the International Nuclear Fuel Cycle Evaluation (INFCE), starting with forty countries and including representatives of fifty-three at the last review meeting in Vienna.

National decisions are, of course, still required as an integral part of international negotiations, and the technological and intellectual leadership of the United States still carries great weight. My strictures against unilateralism do not imply that we have totally lost our leverage over the attitudes and policies of others. Nonetheless, we do need to comprehend sympathetically the assessments and concerns of other nations and to be wary of denouncing as illegitimate or stupid any attitude that differs from our own.

The International Fuel Cycle Evaluation is, in principle, only an exercise in technical evaluation, but much more hangs on its outcome. If it helps to advance toward a genuine global consensus on multinationalization of sensitive phases of the fuel cycle, I see substantial hope for the cause of nonproliferation. If it fails, the outlook is not promising.

REFERENCE

Till, Charles and Y.I. Chang. 1978. "Uranium Implications of Fuel Cycle Selection on Proliferation Grounds." RSS−TM−12. Argonne, Ill.: Argonne National Laboratory, January 1.

Chapter 13

National and International Social Implications

Benjamin D. Zablocki

Every new technology since slash-and-burn has required new social institutions to go along with it, and nuclear technology is no exception. There is, therefore, a need to go beyond Dr. Rowen's very useful and important scheme for decisionmaking among alternative peaceful proliferation schemes. There is a need also to look at the needs for new national and/or transnational institutions that will have to accompany any proliferations in area.

There are five social implications that bear on the need to develop new social institutions. First is the issue of "Great Power" relations, in an era of nuclear proliferation. Second is the conflict between nationalism and internationalism. The third is the issue of the military and diplomatic strategies of small nations, particularly small nations on the threshold of nuclear capacity, and the question of military versus civilian rule in those nations. Fourth, and possibly the most important is the role of multinational corporations in nuclear regulation, and fifth, the question of secrecy and how that bears on power values of primacy in democratic states.

The question of Great Power relations is implicitly founded on the notion of hegemony. The unrecognized problem is that we are in a period of declining American influence over the activities of other nations. Without elaborating the reasons for this declining influence, I will note that it has a great deal to do with the dependence of the United States on materials and the growing interdependence of trade. The possibilities of a trilateral hegemony—comprising the United States, Western Europe and Japan—replacing the American hegemony have been widely discussed and need to be discussed as part of any rational assessment of the possibilities of control of the nuclear proliferation process. It seems doubtful, however, that a trilateral hegemony would have any more ability than the United States alone to influence the nuclear development strategies of the developing nations.

Another possibility is what has been called *pax nordica*, in other words, adding the Soviets to the trilateral nations in order to split the world into an inner ring of northern powers with a monopoly over nuclear technology, and an outer ring of "third world" countries. This would probably solve the problem if the requisite cooperation were available, but at the cost of a highly divisive and undesirable northern/southern split between the "haves" and the "have-nots."

In the absence of credibility for any of these possible strategies for maintaining the hegemony, it seems inevitable that there must be some sort of movement in the direction of a transnational authority, with independent access to force. That such an authority may be the necessary corequisite for a worldwide energy authority with a certain amount of limited sovereignty may not be so far-fetched in the world of the twenty-first century, in the light of the current decline of national loyalties and national legitimacy in all parts of the modern world. In the west, there has been a generational decline in patriotism and the confidence in national governments. At the same time, there has been an increase in cross-national political, religious, and ethnic loyalties, especially among young people.

Small power nationalism should not be taken as a given in the nuclear age. The logic of proliferation of nuclear technologies can be summarized very briefly in the following terms. To paraphrase a statement made by Dr. Gordon "the trade-off for nonproliferation is the even-handed transfer of goods, capital, and technology to the less developed nations." But national loyalties and the legitimacy of loyalty to national regimes is largely founded upon the antithesis of evenhandedness: the quest for comparative national economic advantages. The very existence of nuclear technology may thus be eroding incentives for nationalism in the less developed nations at the same time that is excites motives for nationalistic opportunism.

The third topic that bears on this issue is that of the military and diplomatic strategy of small nations in an era of nuclear possibility. I agree completely with Dr. Gordon's classification of nations at the present time along the dimension of nuclear nervousness, and I agree, also, that it is probably the case that, at any given time, the number of nations sufficiently nervous to motivate them to begin developing nuclear weaponry, may be relatively constant. But I think that point, however well taken, ignores the fact that, over time, access to nuclear weapon technology is a stochastically absorbing state. In other words, nations do not go from having a weapons capability to no longer having a weapons capability, when their international situation improves. It therefore seems to me inevitable that there is going to be steady increase in the number of nations that have nuclear weapon capabilities.

The case of Israeli–Arab relations is instructive in helping us to see how some of these factors may operate in determining the rational and effective strategies of small nations under these conditions. It is, of course, important not to generalize from that very specific and unique case. But the case serves to verify that nuclear capacities can make a difference for small nations even in a world dominated by superpowers. Although nuclear weapons were never used in the conflicts in the Middle East, it cannot be doubted that Israeli nuclear capacity, along with Egypt's singular vulnerability to nuclear attack, provided one of the wedges that finally pried loose the pan-Arab alignment.

To some extent one can argue that effective and continuous superpower vigilance might be effective in preventing this kind of nuclear jockeying for power among the small nations of the world. However, we have not found the superpowers either continuously or effectively vigilant in policing small nation disputes. The fact that the superpowers may sometimes be stalemated, or have more pressing matters to attend to elsewhere, linked with the fact that nuclear weapons capacity is a stochastically absorbing state, leads us inescapably to the assumption that although such conflicts may often be avoided among the smaller nations, there may be times in which they will not be.

If we accept the notion that nuclear threat and nuclear blackmail can be of real benefit to small nations, we are then led to the notion that perhaps democracy and civilian rule may be an evolutionary deficit for the small nations of the world. Again, following the Rowen schema in terms of critical time, secrecy, and so on, it is obvious that civilian governments will be at an extreme disadvantage when compared to centralized military governments, in being able to engage in the necessary kind of diplomacy to accomplish the shift to nuclear weapons capability. But if those nations lacking such capability are thereby placed at a real geopolitical disadvantage, the cause of democracy and civilian rule in the world is measurably weakened. I wish to stress that in terms of this kind of diplomacy there is a very big difference between having no nuclear threat and having a small nuclear threat, especially for small nations (and for terrorist organizations). It becomes very difficult to imagine how any of the strategies for nonproliferation can be very effective over the long run.

We now come to a topic that I consider the most important, although unfortunately it is the one on which I have the least to say. Its absence from the agenda constitutes my major criticism of the way in which this symposium was set up. There exists no expert on multinational corporations at the present conference, and I am dismayed that this is the case. It seems to me that the multinational corporations at the present time constitute at once our best hope and possibly also the chief danger in moving to a nuclear technology without the widespread and possibly disastrous proliferation of weapons capabilities of small nations. Given the unlikelihood that any single nation or group of nations can regulate the spread of nuclear weapons capacity, it seems to me that only the multinational corporations, if anyone, have the grasp of global economic interdependencies to allow them to serve in this capacity.

The obvious objection to global policing of nuclear fuels, "Who will regulate the regulators?" holds *a fortiori* for the multinational corporations. One can easily imagine an effective regulatory role for the multinationals, but one can also imagine situations in which they might succumb to the temptation to go in the other direction and foster the development of nuclear weapons in a small nation. What is urgently needed is a comparative policy analysis of the incentives that may be offered to the multinational corporations to maximize their motivation to discourage nuclear weapons development in the nations whose economies are involved. I do not have a plan for this. All I can say is I hope someone with expertise in this area will think about this matter and come up with such a plan.

My final point has to do with secrecy and the fate of privacy in the transition to nuclear technology under the constraints of having to cope with possible proliferation of nuclear weapons. One of Dr. Rowen's important parameters, a fundamental one, is expressed in "critical time" for the transference from a peaceful use to a weapons use of nuclear technology. Of the many factors of which critical time is a function, I want to highlight one because it bears on this problem.

Critical time is a function of the existing state of the art of surveillance technology as well as of what I call the "Byzantine factor," the capability of any given nation successfully to mobilize large numbers of persons and materials for large-scale and long-term covert activities. An issue that we have to face with regard to nuclear proliferation, if we consider it to be an advantage (and I have argued it is a considerable advantage) for even small nations to have even a minis-cule nuclear weapons capacity, is that those nations that have a highly developed Byzantine factor and have consequently organized them-selves so as to be able to engage the requisite covert activities are going to have a distinct advantage over their other small nation neigh-bors that are more open.

The obstacles to having the Byzantine advantage are such as should give us pause. They include freedom of the press, especially that of investigative reporting; freedom of occupation; freedom of travel; and the clear separation of the civilian from the military sector. One of the costs that we have to consider in assessing nuclear prolifera-tion is the possible continued erosion of democratic institutions in the small nations and the great incentives given to nations to adopt totalitarian regimes.

It seems to me all of this leads to the central implication that the large-scale proliferation of nuclear technology is sociologically in-compatible with the full retention of national sovereignties. However, the lead time for establishing legitimacy for an even very limited global sovereign authority is at least as great as the lead time for developing any new technology. Perhaps, what is worse, it is only within an academic playpen that it is even possible to talk about some of these matters without becoming open to charges of treason!

Discussion

Gene Rochlin: I think that in this whole evening session there has been insufficient separation between the questions of physical security and international safeguards, national and subnational diversion, the difference between a bomb and a threat, and between a known and a potential diversion.

In my mind, things are so tangled up tonight, that the values of various proliferation resistance measures are not adequately characterized. This is more of a complaint than a question.

Marc Roberts: Some of you may know that distinguished American economist, Kenneth Boulding, is fond of writing poetry at conferences. And so with apologies to Kenneth Boulding, I would like to offer the following poem:

> The problem is:
> Who is a wiz—
> Ard enough to follow what's said?
> At this time of night
> I'm not at the height
> Of clarity, inside my head.
>
> The speakers talk fast;
> The diagrams pass,
> As we examine how much Pu is bred.
> "How fast?" someone asks,
> "Can they do the tasks
> That will allow them to make us all dead?"
>
> Yet, all I can think
> Is: "I need a drink!"
> And then, I will stagger to bed.

William Sewell: On that note, I will declare the meeting closed.

Session IV
The Decisionmaking Process and the Plutonium Question

This final session considers decisionmaking about plutonium in the context of national policy decisionmaking in general. The multifaceted character of the policy formation process is illustrated by the variety of disciplines and subdisciplines brought into the discussion. They include management science, macroeconomics, operations research, decision theory, systems analysis, cost/benefit and welfare economics, behavioral science, political science, sociology, anthropology, and psychology. This complexity is a result of the fact that, in the words of Alvin Weinberg, "We're talking about the future; we're talking about world events; we're talking about imponderables." In his opening remarks at the first session Weinberg also remarked, "It will be very interesting to see the degree to which those who hold the most polarized of views can engage in civilized discussion with those at the other end of the spectrum." The two ends of the spectrum in this session turn out to be the managerial-technocratic and the behavioral-humanistic approaches to decisionmaking. After reading the following, the reader can decide whether the discussion, in this session as well as in the three preceding sessions, was civilized.

Session IV

Chairman: *Harvey Brooks*

✳ *Chapter 14*

National Energy Decisionmaking: Rationalism and Rationalization

Marc J. Roberts

This paper is an attempt at "institutional engineering." Unlike many economic analyses of social choice, it does not explore the implications of various hypothetical voting schemes under highly simplified circumstances. (For a more conventional economic approach, see Sen 1970.) Instead, it considers how current social and political arrangements for making energy policy have actually functioned, and how we might make them function more effectively.

Any institutions for making social choices necessarily give more weight to certain kinds of concerns than others. They respond to various kinds of problems in characteristic ways. Alternative sets of institutions thus represent different strategies for solving social problems. The question is: "Can our own current strategy be improved?"

The first two sections of this paper present necessary theoretical preliminaries: first, the kinds of relations we can expect between the structure of institutions on the one hand, and social and political decisions on the other, and second, the criteria we can and should use for choosing among institutional alternatives. The third section reviews what is known about the consequences of current decision-making arrangements. The final section considers the implications of this review and argues for a series of changes in how we organize our affairs.

The discussion leads to several more or less surprising conclusions. First, not only does the United States lack a coherent and consistent energy policy, but we should *not expect* to have one. The way we go about making such decisions is not likely to produce a unified and rational set of programs and practices, especially at the operational level.

Second, current political arrangements overemphasize the selfish and the divisive. Traditional economic interest groups have always played a key role in American politics. More recently, a decline in political parties, the rise of national media, and an educated, mobile electorate have all led to increased influence for ideological interests with strong and clear-cut views. The resulting polarization tends to encourage citizens to act on a narrow view of their stakes in policy decisions, to ignore collective as opposed to particularistic benefits. Current outcomes, I will contend, are too much like what Rousseau characterized in the *Social Contract* as "the will of all" as opposed to "the general will," which he favored.

Finally, what might we do to combat this problem of political particularism? I will argue both for a more effective *managerial* structure in the federal executive and expanded use of *technical* expertise in formulating policy decisions. Such changes should allow us to clarify the nature of the problem and hence to better integrate diverse perspectives in creating and implementing policy. Simultaneously, and perhaps paradoxically, we need to recognize the imperfect, value-laden nature of all "objective" analysis and to make more explicit the role of values in reaching policy recommendations. Only by making more modest claims for expertise can we hope to use it for consensus-building, or even as a framework within which to pursue our disagreements.

DESCRIPTIVE POSSIBILITIES

The relationship among citizen attitudes, institutional arrangements, and social choice is complicated. We should not expect to be able to construct simple syllogisms of the form "All men are mortal. Socrates is a man. Therefore Socrates is mortal." Even looking for a rule (a function) to relate certain characteristics of institutions (independent variables) to specified outcomes (dependent variables) is not sufficiently complex. Rather, we can view institutional arrangements as helping to determine the relationships among the other two sets of variables. For example, expanded rights to a hearing could make it more likely that the views of organized environmental groups will influence certain decisions. But the outcome in any given case will depend on what those views actually are. In technical terms, the regularities we are looking for are not functions, but functionals, that is, generalizations about how relationships vary under different circumstances. (For a discussion of the functional, see, for example,

Allen 1964.) Furthermore, the effect of any one institution will often depend upon how the rest of the system is organized. Thus, it may not be possible to isolate its effect unless we also know about the system in which it is embedded.

What characteristics of our institutions are likely to be important in determining the relevant relationships? The empirical discussion below reflects a conceptual framework I developed and used first in a study of the behavior of six public utilities and later to explore the activities of various regulatory agencies (Roberts and Bluhm 1977; Feldman and Roberts 1978). It seeks to understand an institution's response to external pressures and opportunities by explaining how individual members (and groups) will act, and how these are combined into organizational activity. It amounts to asking, first, what kinds of choices do institutions present to their members? Second, given who these members are and what they believe, how will they respond to those problems of choice? Finally, given the way the institution is organized, how are the choices of their individual members combined into a pattern of institutional action? The resulting way of thinking conforms well to "common sense" notions. It provides a perspective from which to analyze diverse arrangements and a basis for proposing a consistent, specific hypothesis. Although the full development of this perspective is not appropriate here, a brief sketch will help clarify what follows.

External consequences.   We have to begin with the external consequences any organization confronts as a result of behaving in various ways. Will it grow or decline? Whether it is a political party or a federal court, will its members' social status be increased or decreased as a result of various actions?

Individual task definitions and authority.   The decision problem an individual confronts in the organization depends, first of all, on possible choices that he or she can make. What resources are available, given one's formal and informal position in the organization? What decisions must actually be made?

The internal control and incentive system.   For the individual, the consequences of making various choices are a function both of the consequences of such choices to the organization and of the ways in which the organization distributes rewards and punishments to its members. How are people evaluated? For what are they promoted or given other kinds of rewards? How are internal rewards linked to external consequences for the organization as a whole?

Individual preferences and beliefs.   The three previous sets of variables define an individual's decision problem. To understand how a person solves a problem, we must consider that person's values and beliefs and background. What patterns of selective recruitment and socialization are at work? What are the typical educational and class backgrounds of the members?

The mental equipment of real people seldom closely resembles the completely worked out, perfectly consistent preference maps of microeconomics. Instead, individuals have a complex hierarchical structure of both normative beliefs and derived decision rules. Fundamental generalizations and values can be thought of as justifying more detailed descriptive beliefs and normative maxims. An example of the latter is, "Stop at red lights," a rule whose justification involves a whole series of ethical and empirical arguments. Yet, an individual's operational strategies and rules-of-thumb are not necessarily the result of an explicit analysis. Rather, they are often acquired by experience and never fully analyzed—making inconsistent behavior possible, even likely. (See March and Simon 1958 for a classic exposition of such ideas.) What follows is based on this imperfect, problemsolving, rule-based, incremental view of human decisionmaking.

Organization Aggregation Processes

The organization's actions derive from and embody the actions of its members. Most even moderately large organizations comprise a complex internal political system. Various groups and individuals vie for influence and control. The outcome of this competition, as I have elaborated elsewhere (Roberts and Bluhm 1977: ch. 10), depends on various features of the organization's structure, such as who controls task assignments, the nature and length of communication channels, the size and responsibilities of various subunits, patterns of interpersonal influence, and so on.

This framework is only a useful research guide when the organization being studied is only imperfectly coerced by external circumstances. In a highly competitive market, why look inside a firm to explain its behavior, when any departure from profit maximization would lead to its demise? The organization's range of discretion has to be sufficiently large to be worth studying. But that, I would argue, is the case with respect to most current political and social choice institutions. They could respond to citizen attitudes in various ways. To understand the choices that are made requires this kind of internal analysis.

Organizations do evolve over time in response to changing external circumstances. Hence, retrospectively, one can always try to attribute whatever has occurred to those external forces. To do so is incorrect, however. As Lewontin (1978) has recently argued about biological evolution, typically more than one successful adaptation to any situation is possible. The response that is actually chosen depends on the character of the organism (or the organization). This in turn, reflects its prior evolutionary history. In other words, our political institutions could have evolved differently.

Furthermore, the world is not static. Every "ecological niche" is part of a simultaneously determined system in which species, and organizations, define niches for each other. When any one entity responds to the pressures of the larger world, elements of that larger world may react in turn in ways that create still new imperatives. Hence, as economists realize, a "partial equilibrium" look that assumes a stable context may not do. Instead, a "general equilibrium" analysis, which allows for such feedback, may be required. A basic source on such models is Kuenue (1963).

Before exploring such relationships, however, we have to have a better notion of what distinguished desirable from undesirable relationships between citizen views and social policy. Only in that way can we decide what empirical questions we should pursue.

THE NATURE OF
NORMATIVE CONSEQUENCES

As best we can predict, the consequences of any set of institutions will always be uncertain. The "input" variables in the relationships that institutions establish (for example, popular opinion) cannot be perfectly predicted. In addition, institutions do not function with machinelike precision. Hence, the relationships between inputs and outputs are subject to some random variation. Our limited understanding of the system also adds to its apparent unpredictability. As a result, any set of institutional arrangements at best implies a broad distribution of possible outcomes.[a]

Trying to change such arrangements only makes sense if the outcomes associated with various alternatives are sufficiently different. Take, for example, the question of regulatory reform. Trying to alter the whole structure of regulatory arrangements only is sensible if doing so will have more impact than merely making the existing arrangements work better. In any particular case, is the situation more like diagram (a) in Figure 14–1, where most variation is between, not within, institutional arrangements, or more like (b), where the reverse is the case? When outcomes depend more on how

---

[a]Notice that from a decision theory perspective, objective uncertainty (because a system is stochastic) and subjective uncertainty (because we don't know its behavior) are treated equivalently. See Luce and Raiffa 1957.

Figure 14–1.

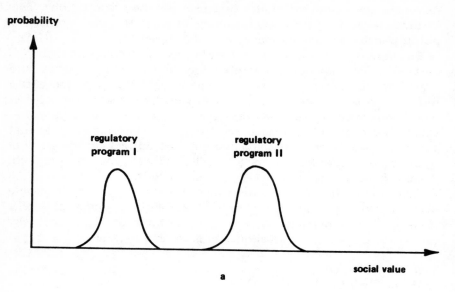

a

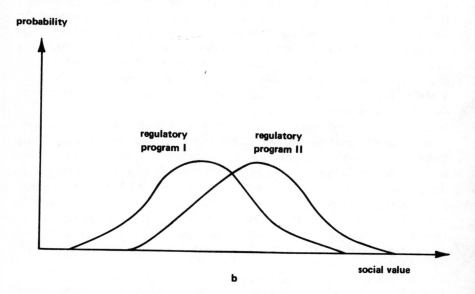

b

institutions are operated than on how they are structured, institutional engineering will not be a wise way to spend time.

The costs as well as the benefits of institutional change are also relevant, particularly when those costs are high. For example, constitutional amendments are difficult to pass. Indeed, one argument for a written constitution is that institutional engineering is not likely to be profitable, and should, therefore, be discouraged by creating a constitution that is hard to change. By this device, everyone's attention is directed instead to the business of running the government. Some institutional changes, like several discussed below, could be implemented by legislative or bureaucratic action. Nevertheless, we still have to decide in each case whether the game is worth the candle.

Designing institutions is more difficult because there is no simple way to compare the desirability of various outcomes. Unlike the diagrams in Figure 14-1, real social outcomes cannot be ranked according to a one-dimensional, quantitative scale. Instead, multidimensional, and often qualitative, evaluations are typically required. One list of possible criteria, taken from an earlier paper, is offered here:

Reliability: The policy should work as intended. The resulting institutions should function adequately in the hands of the kinds of individuals who can actually be expected to operate them. The risk of major departures from planned outcomes should be reasonably low.

Implementation Costs: The costs of gathering the necessary data and of planning and administering the policy (for example, billing and enforcement) should be acceptable given the scale and probable accomplishments of the proposed program. These functions should be appropriately divided between public and private sectors to ensure that accurate and informative data are developed.

Efficiency: The policy should avoid short-run technical and allocational inefficiency. It should respond to differences in the costs of clean-up at different locations and for different sources; to variations in the impact of discharges on ambient conditions, and to variations in the value of improving ambient conditions. All relevant technical alternatives should be employable within the administrative framework.

Stochastic Flexibility: The policy should respond to variations in the state of the ecosystem, to the extent that such flexibility is valuable given its costs and gains.

Dynamic Adaptability: The program should not become entrenched but rather have the capacity to be self-correcting. Some potential for learning and improvement should be designed into the policy.

Technological Implications: The development of new waste treatment technology and new production methods that minimize waste generation should be fostered.

Distribution Equity: The gains and costs of the program, and its net benefits, should be equitably distributed both within and among income, occupation, cultural and geographic groups.

Social and Political Effects: The scheme should foster the development of desirable social and political arrangements and processes. It should be perceived as fair and appropriate, and not injure the vitality of other desirable programs or institutions.

Psychological Impact: Insofar as the program has an effect on tastes for perceptions of the environment, it should serve to make individuals more sophisticated about and better able to appreciate both their options and their own responses to them.

Environmental Risk Aversion: The policy should resolve borderline cases in a manner that tries to minimize major long-term, imperfectly understood, environmental risks.

As this list makes clear, we face decisions of fiendish complexity. There are many dimensions in the "outcome space" and most of these are difficult or impossible actually to measure.

Economists, in contrast, normally evaluate social arrangements by one very simple criterion—economic efficiency (often called "pareto-optimality"). (A clear and critical exposition of this analysis is found in Graff 1967.) Outcomes are divided into two classes. Efficient outcomes are those in which no individual can be made better off without someone else being made worse off. All other outcomes are inefficient.

I find this approach unsatisfactory in several respects. First, it provides only a very weak, two-valued ordering. There is no way to distinguish among various efficient outcomes, nor among various inefficient ones. Yet, the latter is the relevant real-world problem. Second, it involves only a one-time "instantaneous picture," which takes no account of how the system is likely to develop over time. Yet, the dynamic properties of institutional arrangements may be extremely important. Do they generate new information? Do they tend to raise or lower the costs of future change? Those who want to be helpful on real problems have to consider such questions.

The economist's utilitarian perspective is also based on some curious ethical and empirical assumptions. It asks us to judge what occurs entirely in terms of what current individuals currently prefer. This viewpoint tends to be associated with the model of a single, known, and perfectly consistent set of individual preferences noted previously. Much psychological and political theory, in contrast, takes a more complicated view of human motivation. From Plato and Aristotle to Freud and Maslow, many writers have distinguished among various sources of motivation: the higher and lower pleasures, the ego and the id.[b] Such distinctions suggest that human beings do not have a single, well-worked-out preference map. Instead, within the hierarchical structure of beliefs and norms discussed previously, individuals subscribe to conflicting, imperfectly reconciled values. This makes it possible for them to give apparently inconsistent answers to the question of what they "prefer" depending on how the question is posed to them. For example:

1. Would you like the United States of America to give you $100,000 for no particular reason?
2. Would it be a good idea, all things considered, for the United States of America to give you $100,000 for no particular reason?

The fact that many might answer these two questions differently suggests the multiplicity of internal motivations. This same complexity also helps to explain why real people find it so difficult to compare various kinds of outcomes in the way economics says they should. Most people, for example, find it neither easy nor appropriate to answer questions like, "Tell me, how much butter would you trade for your right to vote in presidential primaries?" One need not believe that it is a good thing for people to be inconsistent and limited in their capacities to choose, in order to accept that if they are, we need social choice criteria that reflect this reality. Instead of just trying to satisfy individual tastes, we have to ask what kinds of tastes are being elicited. What sorts of questions do our institutions tend to pose to citizens? What sorts of motivations do they tap, and which perspectives do they elicit as a part of the social choice process?

Real tastes and preferences are also not immutable. Nor are they random accidents or the gift of some indestructible divine spirit. Instead, they reflect each individual's past.[c] Whereas some tastes can be consciously developed (for example, by taking wine tasting or musical appreciation courses), many evolve as by-products of past choices and experiences. And people with different incomes have

---

[b] Indeed, except for the English utilitarians and Skinnerian behaviorists, unitary models of human psychology are quite unusual; see Roberts and Bluhm (1977: Ch. 3).

[c] There is some economic literature dealing with endogenous tastes; see especially Harsanyi (1955: 309); Rothenberg (1953: 885).

often made different choices and had different experiences. In particular, people with expensive tastes usually acquired these as a result of high incomes in the past. Who learns to love yacht racing, for example? Egalitarian utilitarianism thus paradoxically tends to ratify and perpetuate past inequities. Those with expensive tastes, who would otherwise be miserable, can claim a disproportionate share of the world's resources. This helps explain why social policy, as well as some thoughtful observers, have rejected utilitarianism in favor of notions of equal opportunity (as opposed to equal happiness; see Tobin 1970).

The mutability of tastes also means that we should examine the "taste creating" aspects of our arrangements. What sort of *citizens* the state produces can dominate concern over what sort of *goods* it turns out.

Utilitarianism faces still other objections. Almost no one is willing to accept all the implications of the doctrine when confronting nasty or vindictive motives. What do we do about the person who is miserable unless he is making other people miserable? There are other tastes also difficult to accommodate. What do we do about the feeble-minded and the fanatic, and who decides who they are? (One thoughtful discussion of these points is Bergson 1966.)

For these reasons I myself am not a pure utilitarian. I do not accept, except within quite stringent limits, the applicability of the usual social welfare analysis of modern economic theory. And those boundaries are often exceeded when the question is one of institutional design.

But then, what criteria should we use for designing social choice arrangements? Some, discouraged by the complexity of the task, advocate relatively simple decision rules as a way of resolving otherwise apparently insoluble problems.[d] Examples of this are rules of the form, "Pick courses of action that minimize the probability of 'X'," where "X" is some social or physical disaster. Such rules are overly conservative, however. They make it impossible to accept even a very small increase in the probability of such disaster in return for other gains, no matter how large.

Given the nonempirical basis for values, I cannot "prove" that those I will use are "correct." All I can do is try to distinguish among alternatives. Since, as I argued, different institutions alter the relationship between the citizen's values and social outcomes, the relevant question is, "What sort of relationships do we want to create?"

---

[d]Examples of such rules include "maximize biologic diversity" or "minimize the probability of a nuclear accident."

While I believe that all the criteria listed are relevant, there is one especially serious concern in this context, already alluded to, that is critical to the argument.

The issue involves Rousseau's distinction between the "general will" and "the will of all."[e] The "will of all" can be thought of as what people say they want when we ask them questions from the point of view of question (1) posed previously. What would they personally and privately prefer or like, from a selfish perspective? The general will can be thought of as the answer we get to inquiries like question (2). What do people want society and their government to do and be like, both for themselves and others, as a matter of general principle?

Lacking the space to explore this argument fully, I suggest in what follows that we should prefer that kind of democratic society that produces outcomes more along "general will" lines. This is neither a totalitarian nor an elitist suggestion. Both of these "wills" do reflect individual views and values, albeit in different ways. It is simply an acceptance of the importance of civic concern in a world of conflicting values, imperfect decisions, and citizens who are shaped by their own relationships to political process.

THE CHARACTERISTICS OF CURRENT DECISIONMAKING INSTITUTIONS

The problems our institutions are asked to deal with have become both numerous and difficult. Since the New Deal, rising expectations as to the role of government have steadily expanded the public agenda. Increased economic complexity and specialization have produced a society whose members have very diverse views and experiences. We are a long way from the relatively homogeneous commonwealth envisaged by the eighteenth-century liberals. Nowhere is this diversity more evident than with regard to technical knowledge. The sheer expansion in the scope and depth of our understanding has made intellectual specialization unavoidable and the professionalism of expertise widespread. How is democratic government to cope with innumerable issues and divergent interests and to do so when most citizens are inexpert, if not downright uninformed on any given question?

---

[e]"So long as a number of men assembled together regard themselves as forming a single body, they have but one will ... The common good is everywhere plainly in evidence. Upright and simple men are hard to deceive."

There is a classic political science answer to this inquiry. Under such conditions, those groups with the largest, most evident stakes in a decision, which in practice often means economic interest groups, will tend to dominate (Lowi 1969). Such interests have the greatest of incentives to pay the high costs of political participation. They also often have relevant expertise that both lowers the cost of involvement and increases its effectiveness. But recent experience does not fully bear out this prediction, as the following review indicates.

The Legislature

Environmental and consumer legislation—not to mention public action on abortion and gun control—is incomprehensible without recognizing the role of *ideological* interest groups. As swing voters who focus on key issues, their electoral impact may be significant. In addition, they are often willing to pay the high costs of effective lobbying (Wilson 1974). Hence, groups like the Sierra Club or the National Rifle Association can have an impact well beyond their apparent numbers. The growth of such ideological "interests," along with several other social and political developments, have made possible what Ingram (1978) characterizes as "nonincremental" jumps in public policy.

The initiative for such policy departures has often come from subcommittee chairmen in the House and in the Senate who seek national office on the basis of well-publicized hearings and associated legislative initiatives. Estes Kefauver and Crime, George McGovern and Hunger, and Edmund Muskie and The Environment are a few examples. As Ingram (1978) points out, there are costs to such entrepreneurship: friction with one's colleagues, diminished time for attention to constituents, and so on. Hence, only a significant reward—like the presidency—makes such a tactic attractive.

The roots of this pattern lie, in part, in the decline in political parties and the decreasing importance of support from party organizations in primary campaigns. Rising education and geographic mobility among voters have diminished party loyalty and effectiveness. Affluence and the rise of government-provided social services have lessened the demand for party patronage at a time when the professionalization of government leaves it with less patronage to distribute. Simultaneously, the rise in national media gives a candidate direct access to voters and frees him from the need to rely on diminished party resources. These developments make it feasible to cater to geographically diffuse ideological constituencies for the purpose of achieving national media attention.

The preeminent role of committees in the legislative process greatly facilitates this strategy. This role has several causes. First, the scope and complexity of modern government and the arcane technical nature of many issues make it impossible for the typical legislator to be knowledgeable about most issues. In addition, the decline of parties has meant increasing electoral competition in formerly one-party districts. This leads legislators to emphasize attention to constituency needs. With less time to spend on "unimportant" yet complicated legislation, deference to the decisions of a committee only increases. This is especially true in the House, with its shorter terms and relatively small districts.

Within the committee or subcommittee, influence is often concentrated in the hands of the chairman, whose ability to take initiatives has grown with committee staffing levels. Such staff increases reflect several trends. Complaints about how the executive has seized too much of the legislative initiative from the Congress and how the latter was undermanned and lacking in technical expertise have helped. Legislators' own perceptions of the opportunities for advancement from legislative initiatives no doubt also played a role. Typically, the chairman (and ranking minority party member) controls this staff. As a result, minority party members have disproportionate access to the committee's expertise and influence over its legislation writing activities.

Staff increases have enhanced legislative activism in several ways. To justify their existence and provide interesting work, staff members, in turn, have reason to suggest new initiatives. They also tend to be people who lust after the chance to have an impact on legislation. Nor is this a vain hope. Given its expertise, a staff's influence can be considerable.

Within this structure, a few legislators and/or staff members may have a major impact on social policy on a specific question. The role of key staff members in shaping the air and water pollution control acts, for example, is well known (Davies and Davies 1975). As a result, social decision processes can exhibit a high degree of variability. An electoral upset or retirement can lead to major changes in choices and programs as a new chairman and his staff leave their mark.

Despite its variability this process displays some systematic characteristics. Given self-selection, the members of a committee tend to be those whose districts have the biggest economic stake in the subjects under its jurisdiction or those who are eager to exploit its attention-getting potential. Committee staff, likewise, often have either industry ties or idealistic commitments. Thus, within the committee, the strongest economic and/or ideological perspectives tend to be overrepresented. The result is a system that defines and fragments political momentum along geographic and single-issue lines. The obvious, timeworn, and unflattering comparison with the parliamentary system is nevertheless apt in this context.

One consequence of such a system is that we often act before we know very much about a problem. Legislative entrepreneurs first seize initiative in an area. Only then do funds become available to study the problem. By the time that research begins to yield results, the initial program has often become condemned and modified (perhaps more than once).

The simultaneous pursuit of media and legislative politics, in the context of ignorance, often produces what I would call "ambiguous extremism." To attract media attention, the issue is posed in dramatic terms. Sides are chosen, lines drawn, and forceful actions taken. Often the result has been legislative injunctions that mandate Draconian actions on rapid schedules, forbid the balancing of costs and benefits, and promise implausibly rapid solutions. Yet legislative log-rolling also plays a role. To get enough votes for passage, one may need to compromise. This compromise often takes the form of linguistic ambiguity or the inclusion of an exception clause buried in the back of the bill, where only the most dedicated will notice. And, since often no one really knows what to do, the legislation may consist of stern calls for action defined in only the vaguest and most general terms. Agencies are given great discretion and yet are enmeshed in a Rube Goldberg framework of complex and often inconsistent requirements.

The resulting restrictive deadlines and extreme injunctions often just cannot be implemented. The bureaucracy cannot get organized in time. Because the relevant policy questions are often not resolved by the legislation, those implementing the program find it difficult to know exactly what to do. Often, the whole mess winds up in court.

This pattern is not inevitable. Given changing public attitudes, the same structure will not necessarily produce the same outcomes. As voters become skeptical of government, their elected representatives become less eager to be seen as the source of major government initiative. Or the relevant committee leaders may not be in a position to launch a move for national office, because they have to worry about their reelection prospects at home. In such cases, one sees more "typical" incrementalist legislation—in which economic interests dominate. For example, few defend our water resource development policies except the beneficiaries. But recent events show that may be more than sufficient to ensure their continuation.

Regardless of whether economic or ideologic interests dominate—or even where they strike a compromise—the result can be undesirable. The outcome will not be the same as what would emerge if our political institutions, which posed a different set of questions to us, did so in a different way. Combat among the "interests" is not necessarily equivalent to compromise among reasonable and responsible men who are trying to accommodate their differences about a whole range of collective and individual objectives. This is the essence of Rousseau's problem.

The decline of political parties only makes matters worse. Parties can provide some basis for responding to otherwise unorganized interests. Having voters and resources on whose loyalty they can count, parties do not have to always please the most committed on each issue. Any given party has its limits, to be sure. But even the Republican Party's base is broader than that of the Edison Electric Institute or the Environmental Defense Fund. Their decline lessens the capacity of our political institutions for synthesis, for integration, for multi-issue coherence, and for the pursuit of a more or less consistent perspective over time.

The Executive

If Congress is fragmented, regionalist, and particularistic, historically, American institutional engineers have looked to the executive to play an integrative role. Even our political rhetoric reflects this. When disagreeing with Congress, presidents are fond of contending that they alone represent "all the people." Indeed, there may even be something to this image. For the nation's president, history is often the only audience left. This can lead to a largeness of vision and a willingness to take initiatives hard to predict from the pre-election behavior of the individual concerned (Kearns 1977).

Unfortunately, much of the executive branch of government has become as fragmented and particularistic as the Congress. Agencies develop relationships with their respective legislative and appropriations committees (Wilson 1973). In addition, the academic training of an agency's professionals often fosters sympathy with its clientele. The Department of Agriculture is filled with people who know and care about farm problems. The result is a system of mission-oriented "advocacy" agencies—each one defending some narrow interest or concern.

The frequent practice of choosing as agency leader a person who can rally the agency's constituency significantly reinforces this pattern. A "farm state" governor or agricultural school dean becomes the nation's Secretary of Agriculture. A prominent businessman or banker becomes Secretary of Treasury; a labor leader or labor relations academic becomes Secretary of Labor, and so on. Such individuals have connections, ambitions, careers, and preferences that transcend their current jobs. They can have reason to respond to outside interests and pressures. This makes them less than perfect subordinates from the President's viewpoint. But what can he do to them (or for them), except to ask them to resign? And strong constituency support can limit even that mechanism.

The private goals of administrators have varying effects depending on the nature of those ambitions. For example, consider state and federal regulatory commissions. Regulators with political ambitions and a desire for publicity can be self-consciously "tough" and favor ideologic over economic interests. In contrast, those who prefer a quiet life have reason to be industry oriented, to act as if "captured," especially when there is no pressure from the other side (see, for example, Noll 1976).

The structure of the federal administration exacerbates the president's management problems. Dozens and dozens of cabinet and subcabinet officers, regulatory agency heads, and special program heads report directly to him. His ability to supervise, monitor, reward, and punish so many people is very limited. If they were either professionally committed to following orders, or to being personally loyal, this would be less of a handicap. They are often neither. No wonder recent presidents have worked so hard to "control the bureaucracy," and with such limited success.[f]

A president's ability to run the government is further restricted because his subordinates often find it hard to control their own organizations. Many of them are not equipped by training or experience to be effective managers, especially since that job is so difficult. An agency is often staffed by entrenched bureaucrats who joined long before their nominal superior and will serve long after the latter has departed. (The average length of service in a cabinet office has been less than two years in recent times, and in subcabinet offices even less.) Those superiors often do not have the information systems that would allow them to evaluate subordinates' performances. They are also limited by Civil Service regulations in the use they could make of that information, if they did have it. As outsiders, they lack the trust and loyalty of their subordinates, except those they bring with them. And the latter will have the same problems in dealing with those below them. No wonder agencies so often make mission-

---

[f]Consider, for example, the famous example of J.F.K.'s inability to get the missiles removed from Turday, when he so ordered; see R.F. Kennedy (1971).

preserving decisions that minimize change and controversy, despite national policy that points in other directions. (One interesting study in this regard is Pressman and Wildavsky 1973.)

This limited ability of the top to control the middle is especially serious, since in many public activities critical decisions are made fairly far down in the organization. For example, I have argued elsewhere that given limited inspection and enforcement resources, most regulatory activities require a certain amount of voluntary cooperation to be successful (Feldman and Roberts 1978). That, in turn, leads to bargaining between regulator and regulatee over what is to be done—bargaining that is necessarily done by many different people within the agency. Much the same can be said about grant-giving and service-providing activities. When those who actually make these choices are cut off from top-level perspectives and concerns, as they often are, it can be hard indeed to change policy-in-practice or to alter the implementation of a complex program.

When it comes to using this discretion, bureaucracies ironically often choose highly simple, specific decision rules. By avoiding the appearance of discretion, such rules allow one to avoid having to defend individual decisions to highly aggravated parties.[g] Although bureaucrats may be quite content with the appearance of power when handing out goodies when these are in ample supply, the responsibility for denying favors or imposing bans is less eagerly sought.

One response to this inflexibility and uncontrollability has been the extensive development of a nonconstitutional administrative structure. The White House and executive office staff has grown at a great rate to try to compensate for the unreliability (from the president's perspective) of the formal cabinet structure. Personal loyalty to the chief has become a central organizing principle of those staffs. In certain cases, White House staff members have acted as super-cabinet department heads, managing whole areas of government. Cabinet officers, to an extent, report through them. This allows the president to lessen his span of control and deal with a manageably small number of individuals. (Several Watergate-related memoirs offer a clear view of the system at work, including Dean, 1976.) It also helps ensure that someone is, in turn, monitoring the behavior of cabinet officers. (If they, in turn, cannot control their departments, this may not be too useful, however.)

---

[g]For example, early E.P.A. regulations on water pollution control required "secondary treatment or equivalent" everywhere, regardless of costs and benefits; see Roberts (1970).

Even where it lacks implementation authority, the White House staff has become central to legislative development. As a result, within the executive no one person below the president is in a position to consider and integrate diverse ideas from the perspective of longer-run national interests. Instead, the line agencies are left free to be "irresponsible" advocates of their particular constituency and bureaucratic objectives. (For a somewhat similar analysis, see Lowi 1969.) Within the White House staff, on the other hand, the personal ambitions of competing staff members and the president's narrow political interests play a major role. Staff suggestions also tend to be insufficiently influenced by operational considerations since those with such expertise are suspect rivals.

Many other recent institutional developments seem intended to increase the effectiveness and responsiveness of the administrative system and to broaden its perspectives. Notable among these have been various efforts at enhancing "citizen participation." Developments range from requiring hearings or expanding the right to sue, to creating special boards or planning agencies; they represent an implicit criticism of the normal mechanisms of government. Why create special-purpose quasilegislatures with implementation authority unless the existing legislative and executive branches are not doing their jobs? The president's recent proposals for civil service reform and departmental reorganization likewise seem based on a diagnosis similar to the one I have offered. The multiplication of interagency coordinating committees and federal pay raises likewise seem directed at these same problems. Yet support for such measures is limited, as has been their impact to date.

The Judiciary

Another major response to current problems has been to increase the role of the judiciary in social choice. To overcome narrow-minded and fragmented decisionmaking, we have given an increased role to the "philosopher-*Kinder*" of the federal bench. Perhaps these "reasonable men" can strike a balance among various interests, individual and collective, long-run and short?

Antidemocratic elitism is an obvious risk in such a process. Judges are predominantly well off, socially and economically. This tends to influence their sympathies. All the empirical studies, for example, indicate that concern about environmental values increases noticeably with higher income and education. (See the evidence cited in Roberts 1970.)

Judges do face certain incentives—most notably those for legal innovation. At the federal level, most do not plan to run for public office. This frees them from certain short-run political pressures. Yet, except for those on the Supreme Court, advancement is still a possibility. One way to do that is to establish a reputation by writing "important" decisions, which break new ground. Affirming accepted doctrine is unlikely to earn one's work a footnote, still less a prominent place, in a law review article. Furthermore, there are always lawyers and contending parties bringing actions that ask judges to take just such initiatives.[h]

The developments in legal doctrine produced by this process are likely to reflect both judges' personal philosophy and political conditions. (An early exploration of this pattern, which reached slightly different conclusions, is McClosky 1960.) Liberal presidents have often appointed a liberal judiciary that has steadily extended rights, guarantees, protections, and rules. When political opinion swings the other way, judges of a different persuasion tend to be chosen. Once selected, they can express these views, enhance their reputations, and respond to changed incentives, all at the same time, by writing opinions that reinterpret and curtail previous rules. And that, too, we have seen in recent years.

When asked to review agency actions, judges have repeatedly confronted the question of how far to go and what criteria to use. Historic administrative law doctrine restricted the judiciary to reviewing for procedural defects, except for the apparently difficult-to-satisfy claim that the agency had been "arbitrary and capricious." Yet, as these standards have been extended in recent years to whether there is "substantial evidence" to support agency actions, judges have had to look ever more deeply into the substance of agency reasoning (see Stewart 1978).

As Richard Stewart has pointed out (1978: 87–94), this expanded role has its advantages. It has forced agencies to greater self-examination and more explicitness as to the technical bases for their decisions. He notes the evolution of a "paper hearing" that provides for more interchange and examination of scientific and technical materials than is characteristic of traditional "notice and common rule making," but which avoids the rigidities of a full trial-type hearing. This practice has, I agree, been some check to the particularistic and inflexible tendencies of the bureaucracy. Nonetheless, as Sax (1978: 147) points out, we have still not really solved the extremely deep and difficult problem of exactly how much to rely on the peculiar mechanism of an activist judiciary to solve social choice problems.

[h] I am indebted to Steve Thomas for making this point to me.

State and Local Governments

Our discussion, thus far, has focused on the federal level. State and local governments are somewhat different. State legislators, even more than congressmen, tend to be oriented to constituency service. For many legislative service is not their only job and they enjoy only limited staff support. Districts are smaller and personal relationships even more important to political success. States are also more likely to have only a small number of politically important interests, given their lesser economic diversity. Hence, geographically based interest groups (including economic interest groups) are often especially powerful at the state level.

Judicial incentives also may be somewhat different. Some state judges face the same incentives as their federal counterparts, to the extent that they are concerned with establishing a reputation and achieving a higher state or federal post. Historically, however, many members of state courts have had quite different goals. As patronage appointees, they have looked on that role as a terminal position. With few ambitions, they were primarily concerned with avoiding controversy and minimizing personal effort—which produces quite conservative behavior.

As at the federal level, the chief executive of a state is likely to be the point of representation for some less-well-organized interests. Not only does the governor have to run statewide, but governors, like senators, may harbor national political ambitions. These can provoke actions that are hard to account for in terms of intrastate pressures. Dramatically opposing obviously powerful local "interests" is one way to establish a national reputation (see Landy 1975).

On the other hand, at the state and local levels, political competition can act as a restraint on the executive branch. Oster's study of town decisionmaking on sewage treatment plants suggested that plant design was more frugal in towns characterized by greater electoral competition (Oster and Roberts 1973). Incumbents who feared defeat made the effort required to exercise tighter fiscal control. Note that although absolute tax levels are low, relative levels of tax effort were historically very high in poor southern states, which were also characterized by one-party politics.[i]

Given their different political beliefs and institutional arrangements, states do vary considerably in their behavior. In a recent study of the implementation of the Clean Air Act with regard to stationary sources, a coworker and I encountered this quite clearly (Roberts and Farrell 1978). The technical quality of personnel, the

[i]Tax effort in this context is defined as the relationship of average tax revenues to average state personal income.

structure of administrative arrangements, the political climate, and the relationship of the executive to the legislature all played a role. Some states had relatively aggressive programs, and others had relatively ineffectual ones.

Nor have the pressures on the system remained static. At all levels of government, the rise of an organized "public-interest" lobby, particularly in the environmental area, is perhaps the most striking change (Davies and Davies 1975). Now, those having other than economic interests are participating in the process, are knowledgeable about the issues, know the individuals involved, and are a source of data and ideas. This change has contributed greatly to the attractiveness of the issue-oriented entrepreneurship we have observed, which can seem to be (and sometimes is) "progress," versus the old, narrower pattern of influence. Most recently, the countertrend of skepticism toward government may lessen the incentives for legislative entrepreneurship.

What does this review tell us about the acceptability of our institutions and the appropriateness of various reform measures?

WHAT IS TO BE DONE?

The system just described cannot be expected to produce a coherent energy policy. The various committees and agencies participating in the process each respond to their own particular set of conflicting pressures. No one is in a position to ensure consistency of either thought or action. Rather than signifying a lack of will or intelligence, the outcomes we see reflect our highly fragmented, poorly coordinated, special-interest-oriented institutional arrangements. In such a system, when sectional and economic interests have sharply divergent views, it becomes extremely difficult to muster a legislative coalition. Mutual veto and inaction is the likely result. This, then, tends to leave in place whatever hodgepodge of policies has grown up in the past.

In looking toward public action on the plutonium question, therefore, modest expectations are appropriate. A consistent analytically defensible program is not especially likely to emerge. Our institutions were designed to resolve conflict, promote the representation of diverse interests, and respect the views of committed minorities. They were not intended to produce economically efficient schemes of resource allocation. He who designs an elephant has little reason to complain when it fails to fly.

The obvious question is, "Is flying really important?" And, if it is, "Do we really want an elephant after all?" Are there changes we would like to make in the pattern of social choice, and how could we redesign our institutions to achieve such changes? I believe some changes are highly desirable. The current pattern erodes individual commitment to joint ends. Instead of providing incentives and circumstances that foster the development of shared perceptions and common purposes, it legitimizes an attitude of selfishness. "I want mine," becomes everyone's motto for political action. Such an approach tends to produce expansion of the public sector, as no one is in a position to resist a whole series of committed claimants.[j] Eventually, a hostile reaction from otherwise unorganized taxpayers seems likely—as we are now observing. In a world of limited resources and unlimited demands, where relative power is seen as the only basis for resolving disputes, alienation and increasing mutual hostility seem likely outcomes—as recent experience in the United States again all too amply demonstrates.

I believe that such attitudes among citizens, toward each other and toward their institutions, are undesirable. I also assert that "combat among interests" often leads to undesirable outcomes. Depending on who wins, either too much or too little is done from any one of a wide range of middle-ground perspectives. Obvious inefficiencies and socially inequitable patterns of cross-subsidy persist unchallenged— or at least unchanged—because they have support from someone.

Inefficiency, inequity, and ineffectiveness promote dissolution and disaffection of citizens from the political process. Symbolic politics and polarization also foster inflexibility. People strongly committed to moral crusades find incremental adjustments difficult to make. Compromise becomes difficult and current winners fight to avoid reopening each issue—as the politics of natural gas pricing and nuclear power illustrate.

How can we change this situation? How are we to provide more of a role for otherwise unorganized citizens? How can we create an intellectual and institutional context in which at least some disagreements can be resolved in mutually acceptable terms? What role can we find in the political system for something other than narrowly defined interests and relative power? How can we have our political system develop and implement policies and programs in ways that will earn and justify the public's confidence in public actions?

Such questions cannot be avoided simply because they have been debated long and inconclusively by political philosophers of great stature and talent in the past. My own suggestions are not earth-

---

[j] A clear example here is expensive new programs to provide access to public buildings for handicapped. Who will oppose these on cost effectiveness grounds?

shaking. But, then, Utopian visions are not likely to be worth the high—perhaps infinite—costs of implementing them. Instead, I propose attention to two complementary lines of development, one managerial and the other technical, together with increased attention to the limits of these approaches. To use organizational and scientific expertise effectively we must first clarify what it cannot do. For only then can we state its claims in a clear and modest enough manner to have any hope that they will be accepted by the contending parties.

From a managerial perspective, our government is failing to perform certain critical functions that any organization must carry out if it is to respond effectively to a changing world. My own recent research suggests four such tasks (Roberts and Bluhm 1977: ch. 11). An organization must (1) notice changes in the world, (2) develop various options for dealing with those changes, (3) integrate these suggestions into a coherent plan of action that serves the organizations's broad interests, and (4) implement that plan effectively. Our main difficulties come in the last two phases.

Noticing changes in the world requires an organization, at a minimum, to have units or individuals who undertake that monitoring task.[k] Furthermore, such monitors must have enough access to, and influence with, key decisionmakers for their information to be taken seriously. Here, the political system does quite well. Political and media entrepreneurs have every reason to seize on the new and the different. The system will not always respond to change, for reasons discussed, but it is seldom unaware of it.

Second, having noticed a change, a responsive organization has to be able to generate a variety of suggestions and strategies, including unconventional ones, for dealing with it. This means having diverse groups and perspectives represented within the structure. Here, too, the current political system does fairly well. A large national legislature with single-member constituencies produces a very wide variety of viewpoints within the system. Given the media's thirst for "news," on any given question, it is not difficult for a group (or even a well-placed individual) with strong commitments to get his or her proposal "on the table."[l]

The difficulties begin as we move to the third stage. Most of the ideas, suggestions, and plans generated within our political system

---

[k]For example, if a company's R and D effort is to respond to marketing concerns, it needs a strong marketing staff well connected to its research activities; see Lawrence and Lorsch (1967).

[l]Consider, for example, the impact of single academics like Alain Enthoven or Martin Feldstein on health insurance proposals.

have little impact on the decision process. Instead, the committee and agency structure, as we have reviewed, responds disproportionately to a set of relatively narrow interests and perspectives. The integrative process that seeks to find a common plan of action based on common concerns is deficient. And the implementation process, the fourth stage, has parallel difficulties. The lack of effective supervision serves only to further polarize citizens and leads us to overreact to special interests, since they find it easiest to influence the implementation process.

If the political system, considered as an organization, is to improve its capacity for integration and implementation, changes in its structure are required. So, too, are changes in the way explicit strategies are formulated and utilized.

The need for structural change proceeds from realizing that in large organizations that face numerous complex decisions, it is just not possible for top management to perform effectively the entire integrative role itself. The top has not enough time, experience, expertise, or data to do so. Instead, ways have to be found to create "integrators" at lower levels in the organization. One must develop people who are able to, and who have the incentive to, combine the suggestions of various units into the kinds of decisions top management would make—if only it had the time to consider the problem, which it does not.

In addition to creating such integrators, our political institutions require a stronger control and incentive system if there is to be effective and efficient implementation. As in any organization, government employees must face different consequences for taking different actions if they are to have an incentive to behave one way versus another.[m] This requires, as a prerequisite, an information system that allows management to notice variations in performance and a reward system that can be made to respond accordingly. In addition, if top management's control is to reach far down into the organization, so must its power of reward and punishment. Otherwise, unit chiefs dominate the fate of their subordinates, potentially weakening the extent to which members have an incentive to implement what top management intends.[n]

---

[m] A strong control system is thus one in which the conditioned probabilities of rewards for different decisions are quite different. A deep control system is one in which top management can influence those probabilities far down in an organization.

[n] Thus, a strong control system can be "shallow," cutting top management off from middle-level subordinates. Putting promotion power in the hands of group managers obviously reinforces group cohesion, since group managers can now enforce discipline.

But since any incentive system will be imperfect, a well-run organization also requires an explicit strategy. Having such a strategy facilitates implementation since it allows at least some decisionmaking to be delegated to lower levels. It does this by making clear to people what they are supposed to do, and uses their sense of loyalty and professional identity to help achieve conscientious compliance. (No organization exists entirely on incentives, for just this reason; see, for example, Kaufman 1960.) Strategies also make it easier to arrive at most decisions coherently, since first principles do not always have to be reexamined. To be flexible, the organization also clearly requires some mechanisms for strategic reconsideration, a process that can be aided by making sure the strategy is explicit, so that any deficiencies are easier to notice and criticize (Andrews 1971).

How, first of all, are we to build more of an integrative capacity into our choice of political institutions? Integration has often been viewed as preeminently the job of the legislature. However, I believe that contemporary American circumstances require us to look to the executive for performing this function. The legislative alternative would require implausibly drastic steps toward a parliamentary system while the executive is, already, the unit of government most frequently performing this function. Two kinds of executive branch developments seem to be required: creating more of an effective "top management team" to undertake integrative functions and simultaneously finding a way to delegate more authority to reliable "integrators" further down within the structure.

Both of these objectives could be fostered by a significant government reorganization that limited and consolidated existing agencies in new ways. At a minimum, some amalgamation could lead to a series of "super cabinet" positions whose members could serve the functions of, say a "group vice president" of a major conglomerate. Ideally, the occupants of these posts would have two functions. First, they would be responsible for supervising their chief subordinates with regard to agency operations (with appropriate managerial support). They also would play a key role in policy development, serving as the president's key advisors and taking over most of the substantive functions of the White House staff.

For these individuals to function as policy integrators, the units they supervise should be problem-focused. They should include within themselves all the current advocacy agencies that now embody partial perspectives on each major set of issues. Otherwise, the system of interagency bargaining will just continue. What is required is trying to force the integration of the diverse viewpoints on a single problem *within* the boundaries of a single agency. The obvious model is the evolution of the Defense Department out of the previous Army-Navy-Air Force triad—although even here the boundary lines are often illogical and the integration of units concerned with a given mission is very incomplete. Combining Commerce, Labor, and Treasury, or elements of Interior, Energy, and EPA would be other examples. To make such changes most effective, there must be a through intermixing of previous units into entirely new, problem-focused subunits. Then, the leaders of those new problem units could, at a level much below the president, produce policies based on a coherent synthesis of diverse concerns and perspectives; that is, they could serve an integrative function. And their superiors, the new "super cabinet" heads, would be less tied to specific agency positions and constituencies.

The Defense Department example is worth considering in this context. Yes, there is persistent internal rivalry. Still, few would suggest a return to three independent departments. It is clearly valuable to have someone below the President, with line responsibility, worrying about the use of missile submarines versus manned bombers as a way of achieving a given end. The problems in Defense arise, in part, because that situation is especially difficult. Managerially, it involves the most rigid, bureaucratized, isolated personnel system with the strongest distinct professional loyalties of any area in government. Indeed, those personnel systems have been so closed and rigid that top managers have found it difficult or impossible to intervene far enough down within the component organizations to achieve effective control over the main subunits. The typical navy captain or army colonel does not worry about pleasing the Secretary of Defense. In response, McNamara attempted to use formal analysis and a distinct civilian group as an alternative policysetting mechanism, an approach that has not been as successful as some of its enthusiasts hoped. (As an example of McNamara's difficulties, see Art 1968)

This example makes it clear that operating such a system successfully will require complementary changes in the federal personnel system. Increasing performance-oriented rewards and punishments, some of which the current administration has urged, is clearly critical. I believe we must go even further in this regard, however. We must also make some effort at moving middle-level managers among assignments in ways that limit the extent to which they identify with specific units and their clients.

If this management development process is to be carried out, we also have to have top managers in government who serve long enough to be able to get to know their subordinates and to operate the personnel system effectively. This in turn suggests the desirability of a more professional and less political orientation on the part of the cabinet and subcabinet-level offices, as well as a longer term of service. Too many recent occupants of those jobs have had little or no interest, training, or competence in the managerial aspects of their responsibilities.

Not all agglomeration is or will be useful. The inclusion of Education in HEW, for example, represents a much less logical combination than Defense from a problem perspective. That agency is notoriously "unmanaged," in part because no one with managerial skills and interests has been in charge in recent years. And its preexisting structure, too, was left largely untouched until some recent changes. It lacked effective, problem-oriented subunits and instead had functions scattered illogically here and there.

In such a revised context, a greater effort could and should be made to provide a rational management structure within each agency responsible for both the chief operating officer's functions and those of the executive officer. A single top level executive typically cannot do both. Large corporations rely on top management teams for this function, and this logic very much needs to be applied, to a greater extent than it has been, within the federal structure. This is the only way both to provide more emphasis on organizational performance and also to achieve more thoughtful strategy-setting and coordination.

Within a revised structure, efforts should be made to articulate goals and formulate a particular strategy in each problem area. Today, this is often not done, perhaps because advocacy agencies realize that their evident partiality would invite attack if it were too openly articulated. As noted previously, such statements of policy would facilitate delegation, make it easier to judge subordinates, and perhaps serve as a basis for more intelligent congressional review as well.

If this enhanced managerial structure is to be effective, however, it would help greatly to provide a better basis—both within it and vis-à-vis the public—for the conduct of analytic studies of policy options. Today we seem to try to do just that but fail miserably. Ironically, part of the responsibility for this situation lies with the scientists themselves. They are often unwilling to acknowledge the imperfection and limitations of their models. Technicians tend to forget that reaching policy conclusions requires a critical, nonscientific, normative input. But that, too, experts tend to ignore, perhaps because they can claim no special authority in regard to values. Instead of trying to distinguish the roles of fact and value in analysis, and being uncomfortable with emphasizing the latter, scientists often wind up fighting over the "facts,"° even when they are not really at issue. In such a context, expertise itself becomes a commodity for hire.

The relationship of "facts" and "values" is not simple. From a decision theoretic perspective, the "facts" one acts on—for example, which hypothesis about the health effects of a power plant does one use to determine its location—do depend on values. That is, one provisionally accepts a hypothesis based on the probability of making various errors, if you do so, and, the magnitude of those errors. Assessing those magnitudes requires values. What is "objective," at best, is the probability estimates attached to various contingencies.

This same logic applies to the choice of a conceptual framework within which to conduct scientific investigations. Any such framework has certain built-in assumptions about what kinds of evidence are worth considering and what sorts of causal relations are likely to be "important." The choice of certain dependent variables to study involves values about what features of the world are worth trying to explain. All scientific models, like all maps, are partial simplifications. In that sense, all are "false" in some respects. The question is, what formulations are more or less helpful, given one's purposes. Since purposes involve values, so does the choice of what model to use. At least some of our inability to resolve apparently scientific issues related to policy questions has turned on a lack of consensus about what kinds of questions are important and hence what kinds of models and concepts to employ.

The separation of "fact" from "value" can thus only occur in an analytically clear way when scientists become more modest. When doubt occurs about how the world works (as is typically the case when dealing with the scientific issues relevant to decisionmaking about energy), technical experts have to be willing to acknowledge— even quantify—their uncertainty. They have to recognize the role of nonscientific concerns in the choice and use of models. Otherwise, the relevant nature of both policy issues and the extent of scientific disagreement cannot be clearly recognized. If this is done, however, perhaps we can get a more constructive dialogue underway, especially if people come to accept that they need not always claim "different facts" to support an argument, but instead, can rely on explicit value differences. Allowing room for explicit value differences may make a consensus on "facts" easier to obtain.

Of course, the current system does have other virtues. The managerial imperfection of a federal system does allow for experiments, for example, and can increase the political energy of citizens by pro-

---

°One attempt to apply such logic to public activities is *Task Force Hydro.*

viding them with decisionmaking units close to their control. A rational citizen (that is, someone who is "rational" in the narrow economic sense) is more likely to work at politics when the chance of influencing outcomes is larger. And small size could facilitate that. Uniformity can be very inefficient even from an economist's perspective if nonuniformity allows us more closely to satisfy geographically heterogeneous preferences.

I am fully aware of the irony of suggesting a managerial and technocratic solution to a political-theoretic problem. Increasing the effectiveness of the executive branch of government is a two-edged sword, as the recent past demonstrates. I am not, however, simply trying to make the system move more rapidly in response to new problems. On the contrary, some past difficulties have come from doing too much too soon, and too much of the wrong thing, in response to ideological pressures. Yet we live in a world of significant change and great complexity. A political system responsive to well-organized ideologic and economic interests faces obvious difficulties in such a context. Modest institutional changes promise only modest gains, but those seem large enough, relative to their costs, to be worth considering seriously.

REFERENCES

Allen, R.G.D. 1964. *Mathematical Analysis for Economists*. London: Macmillan.

Andrews, Kenneth R. 1971. *The Concept of Corporate Strategy*. Howewood, Ill.: Dow-Jones-Irwin.

Art, Robert J. 1968. *The TFX Decision*. Boston: Little Brown.

Bergson, A. 1966. "On Social Welfare Once More." In *Essays on Welfare Economics*. Cambridge, Mass.: Harvard University Press.

Breyer, Stephen. 1979. "Analyzing Regulatory Failure: Mismatches, less restrictive alternatives and reform." *Harvard Law Review* 92: 549–609.

Davies, Barbara S., and J. Clarence Davies, III. 1975. *The Politics of Pollution*. Chicago: Bobbs-Merrill.

Dean, John. 1976. *Blind Ambition*. New York: Simon and Shuster.

Feldman, P., and M.J. Roberts. 1978. "Magic Bullets and Seven Card Stud: Understanding Health Care Regulation." Paper prepared for the J.F.K. School of Government Conference on Health Care.

Graff, J. de V. 1967. *Theoretical Welfare Economics*. Cambridge: Cambridge University Press.

Harsanyi, J. 1955. "Cardinal Welfare, Individualistic Ethics and Interpersonal Comparisons of Utility." *Journal of Political Economy* 63 (August): 309–51.

Ingram, H. 1978. "The Political Rationality of Innovation." In *Approaches to Controlling Air Pollution*, ed. A. Friedlaender. Cambridge, Mass.: MIT Press.

Kaufman, Herbert. 1960. *The Forest Ranger*. Baltimore: Johns Hopkins Press.

Kearns, D. 1977. *Lyndon Johnson: The Uses of Power.* New York: Basic Books.

Kennedy, R.F. 1971. *Thirteen Days: A Memoir of the Cuban Missile Crisis,* Richard Neustadt and Graham Allison, eds. New York: Norton.

Kuenue, R.E. 1963. *The Theory of General Economic Equilibrium.* Princeton, N.J.: Princeton University Press.

Landy, Marc. 1975. "The Regulation of Strip Mining in Kentucky." Ph.D. dissertation, Department of Government, Harvard University.

Lawrence, Paul R. and Jay W. Lorsch. 1967. *Organization and Environment.* Cambridge, Mass.: Graduate School of Business Administration.

Lewontin, R.C. 1978. "Adaptation." *Scientific American* 239 (September): 212–30.

Lowi, T.J. 1969. *The End of Liberalism.* New York: Norton.

Luce, R. Duncan, and Howard Raiffa. 1957. *Games and Decisions.* New York: Wiley.

March, J.G., and H.A. Simon. 1958. *Organizations.* New York: John Wiley.

McCloskey, Robert J. 1960. *American Supreme Court.* Chicago: University of Chicago Press.

Noll, Roger G. "The Consequences of Public Utility Regulation of Hospitals." In *Controls on Health Care,* Papers of Conference on Regulation in the Health Industry, January 7–9, 1974, pp. 25–48. WDC: National Academy of Sciences, Institute of Medicine, 1975.

Oster, S., and M.J. Roberts. 1973. "Water Pollution Control and Political Process: The Merrimack Valley." Unpublished.

Pressman, Jeffrey, and Aaron Wildavsky. 1973. *Implementation.* Berkeley: University of California Press.

Roberts, M.J. 1970. "River Basin Authorities." *Harvard Law Review* 83, no. 7 (May): 1527–56.

Roberts, M.J., and J. Bluhm. 1977. "The Reponsive Organization." Unpublished.

Roberts, M.J., and S.O. Farrell. 1978. "The Political Economy of Implementation." In *Approaches to Controlling Air Pollution.* ed. A. Friedlaender. Cambridge, Mass.: MIT Press.

Rothenberg, J. 1953. "Welfare Comparisons and Changes in Tastes." *American Economic Review* 43 (December): 885–90.

Sax, Joseph L. 1978. "Comment." In *Approaches to Controlling Air Pollution,* ed. A. Friedlaender. Cambridge, Mass.: MIT Press.

Sen, Amartya K. 1970. *Collective Choice and Social Welfare.* San Francisco: Holden-Day.

Stewart, Richard. 1978. "Judging Imponderables of Environmental Policy." In *Approaches to Controlling Air Pollution,* ed. A. Friedlaender. Cambridge, Mass.: MIT Press.

Tobin, J. 1970. "On Limiting the Domain of Inequality." *Journal of Law and Economics* 13, no. 2 (October): 263–277.

Wilson, J.Q. 1973. *Political Organization.* New York: Basic Books.

_____. 1975. "The Politics of Regulation." In *Social Responsibility and the Business Predicament,* ed. James W. McKie. Washington, D.C.: The Brookings Institution.

Chapter 15

Probability Assessment and Decision Analysis of Alternative Nuclear Fuel Cycles*

*Alan S. Manne and
Richard G. Richels*

This paper reports upon a decision analysis experiment conducted among sixteen individuals who were well informed on the topic of nuclear weapons proliferation and on the civilian use of plutonium fuel cycles. The results are troublesome to those who believe in the possibility of objective scientific judgments in an area as polarized as this one has become.

Depending upon whether one supports or opposes plutonium fuel cycles, one also tends to adopt arguments on uranium availability, energy demand growth, and so on that support one or the other position. This finding could help explain why it has proved so difficult to reach a factual consensus within expert working groups, each specializing in a particular aspect of the overall problem. Consciously or subconsciously, it appears that the expert committees are simultaneously debating the larger issue that confronts society.

AN INTRODUCTION TO DECISION ANALYSIS

Decision analysis represents a promising approach for dealing with controversial policy issues. It is a pragmatic technique for comparing alternative futures, for it focuses attention on which actions need to be taken today and which ones can be deferred pending the resolu-

*This paper is a revised version of the paper delivered at the symposium. The authors are indebted to Hung-po Chao for his assistance with these calculations. They are also indebted to Henry Rowen for helping to coordinate their research with that conducted elsewhere within the U.S. government's Nonproliferation Alternative Systems Assessment Program (NASAP).

tion of uncertainties. It provides a way to combine the strengths of two ancient maxims: "Look before you leap"; "You can cross that bridge when you come to it."

In principle, the approach is straightforward. Using a tree diagram, the problem is described as a series of decision and chance nodes unfolding sequentially over time (for an exposition of this approach, see Raiffa 1968). Through this process, policymakers are helped to define a set of feasible alternatives, uncertainties, dates of resolution of uncertainties, and the possible outcomes. *If* there is agreement on some such goal as economic efficiency, the approach can be employed to identify an optimal strategy: one that maximizes the expected net benefits, allowing both for upside and downside losses.

Here is a simple hypothetical example for a yes–no decision on R&D (research and development) expenditures proposed for federal funding to help develop a new energy supply or conservation technology. In designing this program, one must attempt to balance two types of economic costs, each evaluated in terms of discounted "present" values:[a] (1) a *downside loss* that, say, $5 billion of R&D costs will have been incurred unnecessarily if the technology is developed and then—for one or another reason—it is not brought into widespread commercial use; and (2) an *upside loss* that, say, $100 billion of costs will be incurred if the technology is not available at a future date when other energy resources become even more expensive than they are today. This upside loss is sometimes termed the "gross economic benefits" of developing the technology—that is, before netting out the R&D costs.

Provided that the upside and downside cost estimates summarize the relevant social and private aspects of this R&D proposal, and also provided that there is agreement on the *probability* that the technology will be brought into widespread commercial use, it is clear that the R&D funds should be expended if and only if:

R&D costs (downside loss)	$\leq$	$\begin{pmatrix} \text{probability that} \\ \text{technology will} \\ \text{be brought into} \\ \text{commercial use} \end{pmatrix}$	$\begin{pmatrix} \text{discounted gross} \\ \text{economic benefits} \\ \text{from commercial use} \\ \text{(upside loss)} \end{pmatrix}$
$5 billion	$\leq$	(p)	($100 billion) .

[a] Suppose that we have alternative ways of investing money that will yield 5 percent per year. Then if someone promises to pay us $1000 twenty-five years from today, the "present worth" or "discounted value" of this asset is said to be $295. This is the amount which would accumulate to $1000 in twenty-five years at a 5 percent rate of return on investment. In general, if the benefit is not realized until t years from today, and if we have alternative investment oppor-

In this example, note that the breakeven value of p is 0.05. That is, it would be worthwhile to develop this technology if the chances are better than 1/20 that it will eventually be used.

The principal objections to this type of analysis usually take three forms. (1) It is exceedingly difficult to compare conventional economic costs and benefits with possible social consequences such as nuclear weapons proliferation that may be associated with civilian plutonium fuel cycles. (2) It is difficult or impossible to reach an objective agreement on probability assessments such as the parameter p. And (3), it is difficult or impossible to agree on a public discount rate for comparing today's R&D expenditures with the twenty-first-century economic benefits that may be provided by the new technology under consideration. For the balance of this paper, we shall concentrate on the first two of these barriers to the use of decision analysis.

On the rate of discount to be applied by the federal government, there is a longstanding debate within the economics profession. Some have urged that a zero rate be employed, but others have noted that the United States simply could not afford to finance the large number of projects that would then become economically justifiable. Even if we reject a zero discount rate, this does not mean, however, that it is appropriate to base long-range R&D decisions upon the current guidelines of the U.S. Office of Management and Budget: a 10 percent annual discount rate, measured in dollars of constant purchasing power. A 10 percent rate may be appropriate for federal investments competitive with those of private enterprises that are required to pay state and federal taxes, but 10 percent is a good deal higher than the real (inflation-corrected) after-tax returns on capital that are available to most investors in the United States today. As a compromise criterion, we shall therefore apply a 5 percent real discount rate to future public costs and benefits. It is assumed, however, that private producers and consumers will continue to base their decisions on capital charge rates such as 10 percent per year— including the effect of corporate and personal income taxes on their investment and pricing decisions.

ALTERNATIVE NUCLEAR FUEL CYCLES

We shall now turn to the more emotionally charged issue of nuclear fuel cycles, considering: (1) a comparison of the economic costs and

---

tunities available yielding r percent per annum, the present value would be $[1 + (r/100)]^{-t}$. Throughout, we shall calculate the discounted value of costs and benefits in terms of dollars of constant 1975 purchasing power—rather than attempting to forecast the future of U.S. inflation.

benefits with the geopolitical consequences of proliferation/diversion, and (2) the objectivity of individual probability assessments. Probability assessments will be reported for a group of sixteen well-informed individuals (from U.S. government agencies, universities, industry, and environmental defense groups), all of whom were participating in a seminar conducted for the Nonproliferation Alternative Systems Assessment Program (NASAP). This paper will be restricted to the evaluation of alternative nuclear fuel cycles. We will not attempt to deal with a more general issue: the costs and benefits of terminating the development of civilian nuclear power altogether, including that based on today's conventional LWR (light water reactor) with a once-through cycle. Clearly there are uncertainties concerning reactor safety, spent fuel disposal, and proliferation/diversion for the once-through LWR, but these are not the principal focus of this paper.

Before proceeding further, it may be worthwhile to summarize the key features of these alternative fuel cycles. The plutonium-fueled FBR (fast breeder reactor) transmutes fertile into fissile isotopes. It is said to "breed" because it produces more fissile material than it consumes, hence requires little or no input of natural uranium fuel. The excess fissile material (plutonium, etc.) produced by an FBR may be recovered from the spent reactor fuel through a chemical "reprocessing" plant, and may then be recycled back into an FBR or into other nuclear reactors. Unless special protective measures are taken, weapons-usable material could be diverted either from an FBR or from a reprocessing plant. In the case of today's once-through LWR fuel cycle, the diversion of plutonium would be considerably more difficult.

An LWR is said to be a "converter" reactor because it converts only a fraction of its fissile material into additional fuel. An advanced converter reactor (ACR) would have a higher conversion ratio than the LWR, hence requiring less natural uranium input per KWH of electricity produced. Some improvements in the conversion ratio may be obtained with once-through fuel cycles (based on light or heavy water or on gas-cooled reactors). For high conversion efficiencies, however, a chemical reprocessing step would be needed. There is a wide range of opinions on the relative difficulty of diverting weapons-usable material from the once-through LWR, the ACR and the FBR.

THE TECHNOECONOMIC UNCERTAINTIES

During the past decade, there have been enough benefit-cost analyses of alternative nuclear fuel cycles so that we encountered little difficulty in drawing up a *list* of the key technoeconomic uncertainties affecting this issue. (For a review of earlier cost-benefit analyses, see Manne and Richels 1978, which also summarizes the ETA–MACRO model of energy-economy interactions.) This is sometimes termed "second-order agreement." That is, individuals will differ in their numerical estimates of the quantity of uranium resources available or on the feasibility of energy conservation, but there is no debate that these are among the central issues that must be confronted when evaluating alternatives to the plutonium-fueled FBR.[b] After some preliminary experimentation, we settled on the following short list of technoeconomic uncertainties. These uncertainties affect primarily the gross economic benefits of developing one or another nuclear fuel cycle; they were the first seven issues to be considered by the participants in our probability poll:

1. The ease of energy conservation (as measured by the "elasticity of substitution" between capital, labor and energy);
2. The cost of distant-future alternatives to oil and gas (such as solar heating, biomass, or unconventional petroleum resources); all these options are summarized by the abbreviation AES, nonelectric alternative energy systems;
3. Environmental and institutional constraints on coal utilization;
4. Political and regulatory constraints on the expansion of once-through LWR capacity;
5. The dates and rates of introduction and the uranium consumption requirements of advanced converter reactors (abbreviated ACR); it has been argued that an ACR could be developed so as to be more uranium-efficient than today's LWR, and also less susceptible to proliferation/diversion than the FBR (the ACR has been advocated as an alternative to the FBR by Feiveson, von Hippel, and Williams 1978);
6. The magnitude of uranium resources available at alternative prices;
7. The cost of advanced electric technologies (here abbreviated ADV). This category includes clean, safe, inexhaustible and as yet *un*developed electric energy sources such as solar, fusion, and hybrids.

In arriving at our assumptions on costs and reactor characteristics, we relied heavily upon the preliminary reports that had been assem-

[b]Prior to the Carter administration, the FBR had been central to the nuclear reactor strategy of the United States and other industrialized nations. For an early critique of the FBR, see Cochran (1974). For a more recent critique, see Kenny et al. (1977).

bled by the national laboratories for use by NASAP. For an overview of these estimates, see the levelized busbar costs shown in Figure 15—1 and Table 15—1. This provides immediate insights into the possible magnitude of the gross economic benefits. When uranium costs $50 per pound, note that the LWR holds a slight advantage over coal-fired units in some regions of the United States and that the reverse is true in other regions. That is, on NASAP assumptions, the LWR has a significant advantage over coal only in the event that there are environmental or other constraints that prevent, say, a quintupling of today's rate of U.S. coal consumption.

From Figure 15—1, the reader will note that the once-through LWR is at least as economical as the FBR and the two alternative types of ACR, provided that uranium prices remain at $50 per pound. It is only when uranium resources become depleted and prices triple to $150 that these other reactor types become significantly more attractive than the once-through LWR. Depending on the magnitude of uranium resources, the rate of growth of electricity demand, and so on, the ACR could fit into a "time window" before it becomes desirable to commercialize inexhaustible electric energy sources such as the FBR and the ADV.

The width of the ACR's time window also depends upon the *date* at which it could be commercialized. Our base case assumption is that 2010 is the earliest date at which there could be 6 GW of FBR or of either type of ACR. (The ACRA and ACRB differ from each other in their uranium consumption and in their fuel reprocessing requirements.) There is little dispute that by 2010, one could commercialize an ACR consuming only 96 short tons of uranium per GW-year (versus 165 short tons for today's once-through LWR). It is arguable, however, whether the ACRB technology could be developed by 2010—or even by 2020. This is one of the uncertainties posed to the participants in our probability assessment poll.

The ADV (solar, fusion, or hybrid) presents another uncertainty with respect to timing. For these calculations, we have supposed that a low-cost ADV (one with electricity costs comparable to those of the FBR) could be developed in any event by 2030. The principal uncertainty is whether an accelerated R&D program could achieve this low a cost level by 2010, or whether the costs of an early ADV are likely to be much higher, for example, 2.5 times the level now anticipated for the LWR.

In each case, we have supposed that there are constraints upon the speed of adoption of these technologies, and that their growth rates cannot exceed those that had been optimistically anticipated for the development of the LWR just after it had reached the 6 GW level in

Figure 15–1. Levelized Busbar Cost Assumptions

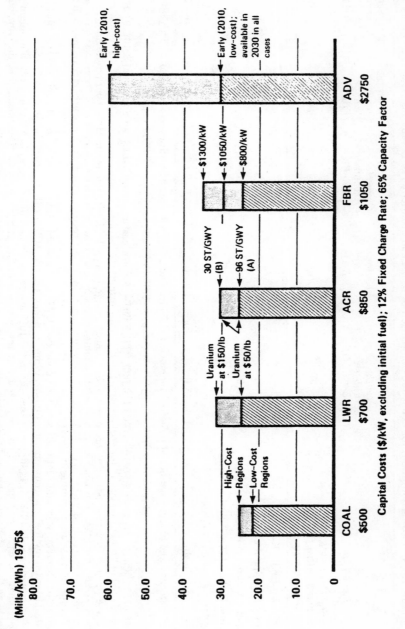

(Mills/kWh) 1975$

Capital Costs ($/kW, excluding initial fuel); 12% Fixed Charge Rate; 65% Capacity Factor

Table 15–1. Cost Assumptions for Economic Comparison of Base-Load Electricity Plants. (1975 General Price Level)

Plant Type	Coal-Fired (with Scrubbers) (COLL-COLH)	Light Water Reactor, No Recycle (LWRA)	Advanced Converter Reactors (ACRA-B)	Fast Breeder Reactor (FBRX)	Advanced Electric, Early (ADVA)
Capital Costs ($/KW) (excluding initial fuel inventories), cap_i	500	700	850	1050	1375–2750
Unit cost (mills/KWH):					
Levelized capital costs[a]	10.6	14.8	17.9	22.2	29.0–58.0
Coal/uranium resource costs	8.0[b]	3.9[c]	2.0–1.8[d]	0	0
Other fuel costs	0–4.0[b]	3.3[c]	3.6–8.8[d]	5.4[e]	0
Operating and maintenance	3.0	2.0	2.0	2.0	1.0–2.0
Total busbar costs	21.6–25.6	24.0	25.5–30.5	29.6	30.0[f]–60.0

[a] 12 percent fixed charge rate, net of inflation. Includes taxes, insurance, depreciation and utilities' cost of capital. Therefore:

$$\text{levelized mills/KWH} = (0.12/\text{year}) \, [cap_i \div (0.65 \, (8.76 \; 10^3 \; \text{hrs/yr}))] = 0.0211 \; cap_i.$$

[b] $8.0/10^3$ KWH = (coal at $.80/10^6$ BTU) (10,000 BTU/KWH)—for low-cost regions (COLL); an additional $.40/10^6$ BTU is included in "other fuel costs" for coal transport to high-cost regions (COLH).

[c] These nuclear fuel cost estimates are based upon a once-through fuel cycle (LWRA), uranium at $50 per pound and enrichment at $85 per SWU. For simplicity, plutonium credits are taken to be zero in this table. A more realistic evaluation of plutonium is incorporated within ETA-MACRO. Our calculations allow for 1 percent annual evolutionary improvements in the LWR for thirty years beginning in 1985.

[d] Reactor characteristics and fuel costs are based upon two HWR cycles described by Feiveson, von Hippel, and Williams. Our calculations allow for 1 percent annual evolutionary improvements in the ACR for thirty years beginning in 2020.

[e] For an early oxide FBR, it is supposed that the initial fuel inventory (in and out of core) is 5.0 tons fissile plutonium per GWe, that the capacity factor is 65 percent, and that the annual breeding gain is 4 percent. This means that the compound doubling time is eighteen years, and the simple doubling time is twenty-five years.

[f] Costs of *late* advanced electric (ADVX) are identical to those of low-cost ADVA: 30 mills/KWH. The ADVX technology is not available until 2030.

1970. This means that it might be possible to grow at a 40 percent annual rate during the first five years after reaching 6 GW, at a 30 percent annual rate during the second five years, 20 percent during the next five years, and up to 10 percent annually thereafter. Because of these limits on market penetration, it turns out that the United States could be deploying additional LWR capacity *after* the year 2010, even if the FBR or ACR were commercially available at a 6-GW level by that date. For an example of such a scenario, see Figure 15−2(a). This is based upon the following resolution of the key technoeconomic uncertainties:

1. Elasticity of substitution between energy and capital-labor = 0.25
2. Nonelectric AES cost = $6/MMBTU
3. Coal constraints: 40 quads in 2000; 60 quads asymptote
4. Limit on LWR capacity in 2000: 300 GW
5. No ACR; 6 GW of FBR in 2010
6. Amount of natural uranium available at $150/pound: 3.0 million short tons
7. Cost of early advanced electric: 2.5 × LWR costs

GROSS ECONOMIC BENEFITS— ALTERNATIVE SCENARIOS

Figure 15−2 (and all other scenarios presented here) is based upon the ETA-MACRO model of energy-economy interactions. This allows for price-induced energy conservation (for example, the use of insulation to replace heating fuels in homes and other structures), and also for interfuel substitution (such as the use of electrically driven heat pumps in place of oil and gas burners). The model incorporates the effect of rising energy prices upon capital formation, hence slowing down economic growth. In addition to the key uncertainties just listed, the following assumptions underlie these calculations:

• At constant energy prices from 1970 to 2000, the *potential* rate of GNP growth would be 3.4 percent per year. (Because of energy price increases, however, it turns out that the *realized* growth rates are of the order of 3.0 percent.) After 2010, the anticipated slowdown in the labor force and productivity limits GNP growth to only 2.0 percent per year—well below the rates experienced over the period 1950−70.

Figure 15-2. Energy Supplies and Demands

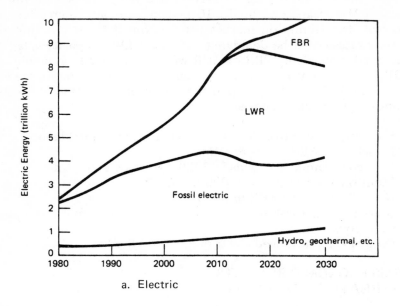

a. Electric

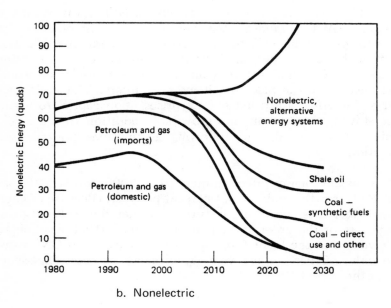

b. Nonelectric

- Price controls on domestically produced oil and gas will be gradually relaxed so that domestic consumers pay the incremental replacement costs of energy.

- Domestic oil and gas resources are 1600 quads recoverable at prices of $2—$5/million BTU. With this resource base, it is possible that domestic oil and gas production could rise at 1 percent per year from 1980 to 1990.

- OPEC will pursue a cautious and surprise-free pricing and production policy. Given the high costs of alternative international sources of supply, OPEC will be able to raise prices at an average of 4 percent per year (in real terms) after 1980.

Figure 15—2(b) shows the implications of these assumptions. Note the slight rise and then the gradual decline of domestic oil and gas production. Rising prices constrain total demand for nonelectric energy, but imports continue to play a major role until after the year 2000. It is not until then that there is a significant economic incentive to introduce alternative domestic sources: shale oil, coal-based synthetics, and the AES. At that point, there is also a significant incentive to substitute electricity for oil and gas. By 2020, 53 percent of primary energy inputs are converted into electricity—in the scenario described by Figures 15—2(a) and (b).

To employ these scenarios within a decision analysis framework for alternative nuclear fuel cycles, each case must be run twice, first with and then without the specific technology introduced at a given date. The *difference* in the present value of the economy's overall consumption level is then reported as the gross economic benefit. Under the assumptions that underlie Figures 15—2(a) and (b), for example, it makes a difference of $200 billion (in present value) whether or not the FBR and plutonium recycling are introduced. These potential benefits are a good deal larger than the RD&D (research, development and demonstration) costs, but represent less than 1.0 percent of the twenty-first century's discounted GNP. (Compare leftmost bar with the two rightmost bars in Figure 15—3.) Under other assumptions, the gross economic benefits of the FBR could be lower—or zero if uranium prices fail to rise.

In all cases that we have examined, the ACR yields lower gross economic benefits than the FBR *when they are introduced at the same date.* Under the circumstances that make the FBR economically justifiable (low uranium supplies relative to the demand for nuclear energy), there is only a narrow time window when the ACR is more attractive than its competitors: the ADV and the LWR oper-

Figure 15-3.   Present Value of Costs and Benefits

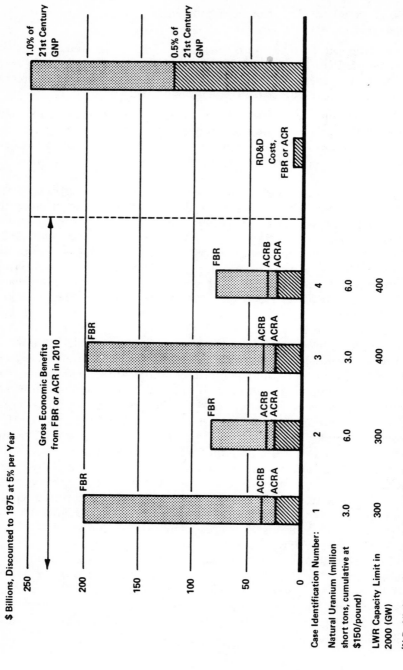

ating on a once-through fuel cycle. Under other circumstances (high uranium supplies), there is no economic incentive to commercialize either the FBR or the ACR. The ACR would be advantageous only if it could be introduced earlier than the FBR—*or* if it had a significantly lower potential for proliferation/diversion. We shall now examine these issues in turn.

ALTERNATIVE R&D TIMETABLES

Figure 15–4 summarizes five broad alternatives for reactor and fuel cycle development in the United States today: parallel development of both an FBR and ACR demonstration reactor, development of just one type, wait, or stop. The squares denote decision points in 1990 and 2000. Many of today's uncertainties will have been resolved by these dates. Some speedup (and perhaps a reduction in R&D costs) might be achieved if this program were undertaken cooperatively with Canada, Japan, and the Western European countries. (On the prospects for such cooperation, see the Rockefeller Corporation 1978.)

These specific timetables are based upon *sequential* development of the initial demonstration and commercial-scale plants in the United States. This would provide a maximum opportunity for learning from one stage to the next so as to avoid repeating the mistakes that are inevitable in any large-scale development program. Under sequential development, note that 2010 is the earliest plausible date for 6 GW of either FBR or ACR prototype capacity. If, for example, an FBR demonstration plant were authorized today (or by 1980), this unit could be completed before a 1990 decision were required on its possible successor—here labeled CBR–1, the first commercial-scale breeder reactor. In turn, CBR–1 could be completed by 2000— prior to the date at which decisions were required to authorize the balance of the 6 GW of prototype capacity to be available in 2010.

In the case of the ACR, it might be possible to take advantage of existing Canadian experience with heavy water reactors, but there would still be an extensive program required to develop a thorium-uranium–233 isotopically denatured fuel cycle. All told, this means that there could be only a minor difference between the research, development, and demonstration costs of the FBR and the ACR. Our specific numerical assumptions are shown along each arc of Figure 15–4. Any of these values could be changed substantially without affecting the overall conclusions. Note that the "wait" strategy would lower the present value of the R&D costs. By delaying the date for reaching 6 GW of capacity, however, the "wait" strategy also lowers the potential gross economic benefits.

Figure 15–4.   RD&D (Research, Development, and Demonstration) Cost Summary ($ billions, discounted to 1975 at 5 percent per year). N.B. Number on each arc denotes present value of RD&D costs for that segment of the program.

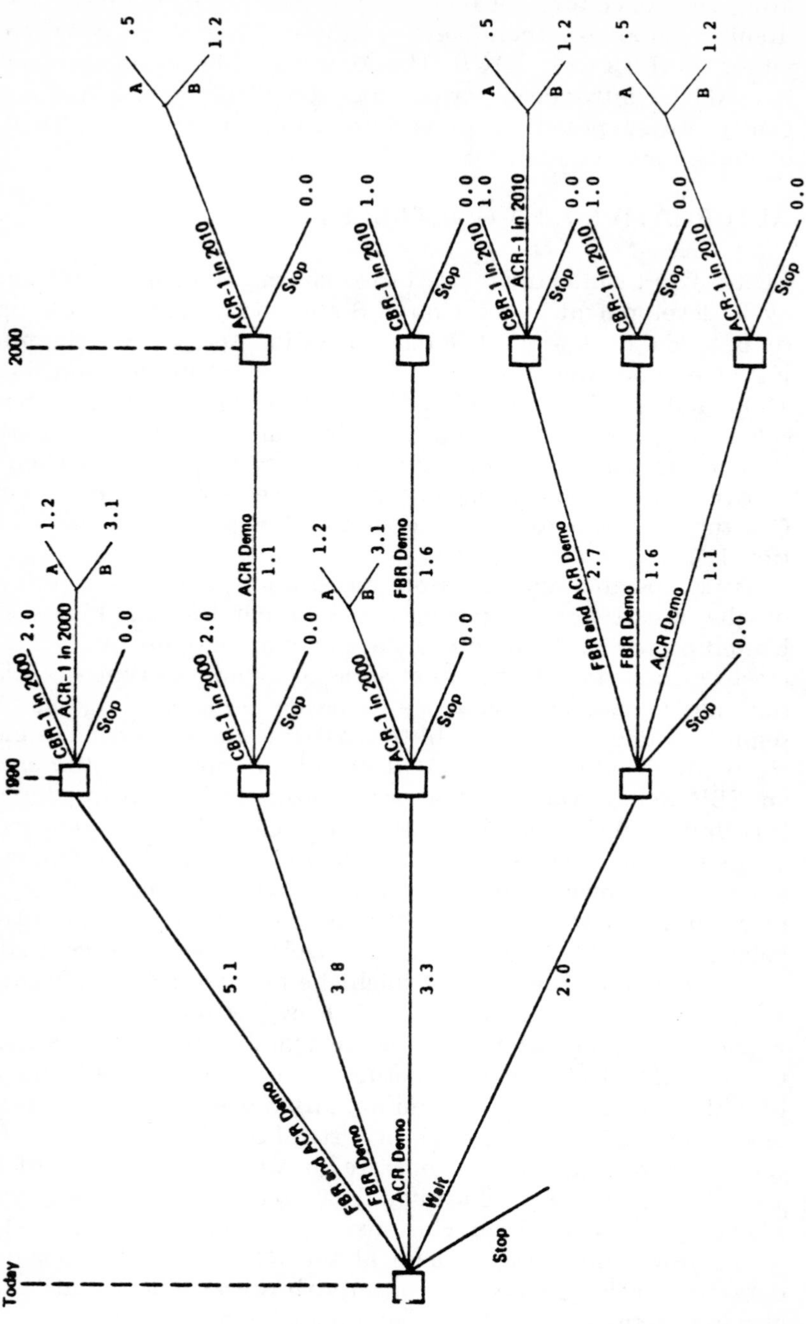

PROLIFERATION/DIVERSION RISKS
OF ALTERNATIVE FUEL CYCLES

On the geopolitical risks of proliferation/diversion, there is even less objective evidence available than on the technoeconomic uncertainties. Rather than attempt to review the entire proliferation controversy, we shall begin by quoting two alternative viewpoints. The first represents a brief summary of what later became the Carter administration's position on plutonium fuel cycles: "The risks associated with reprocessing and recycle of plutonium weigh strongly against their introduction. The use of plutonium in the commercial fuel cycle would expose to diversion and theft material directly usable for weapons" (Keeny et al. 1977: 30). The second quote comes close to expressing the European and Japanese governments' reasons for opposing the Carter administration on this issue: "In brief, it seems to us that if the U.S. were to forego the option of expanding its nuclear energy supply, the global scarcity of usable energy resources would force other countries to opt even more vigorously for nuclear power, and moreover to do so in ways that would tend to be internationally destabilizing" (Rose and Lester 1978: 45).

To summarize: according to the first quotation, the commercialization of plutonium fuel cycles within the United States could make a significant difference in the number of countries that are willing and able to join the six that now belong overtly to the nuclear weapons club: the United States, the Soviet Union, the United Kingdom, France, China, and India. The United States should therefore be willing to forego a fraction of its GNP rather than see nuclear weapons spread into additional hands. How large a fraction would it make sense to give up? Clearly this fraction would have to be a good deal lower than 100 percent of the GNP. For orientation, it is worth recalling that 1 percent of today's GNP represents $20 billions per year, that our total military defense budget is 6 percent of the GNP, and that our total official aid to developing countries has dropped to 0.25 percent of the GNP. These are the percentages that should be compared with the possible impacts upon proliferation/diversion.

According to the second quotation, the commercialization of plutonium fuel cycles within the United States would reduce our future need to import oil and uranium—hence lower the pressures upon limited worldwide energy resources. This would reduce international tensions, and so the international impacts should be counted as a benefit rather than a cost to the United States. Still a third position is of course possible: that the geopolitical impacts are virtually negligible.

Table 15–2.   Results of Poll Taken at NASAP Summer Study Meeting, August 15, 1978

Date of Resolution	Resolution	Respondent	1	2	3	4
1. 1990	1. Elasticity of substitution between energy and capital labor	0.25	0.9	0.75	0.2	0.67
		0.50	0.1	0.25	0.8	0.33
	(Measures percent reduction in primary energy demand for a one percent increase in relative price of energy.)					
2. 1990	2. Cost of nonelectric AES in 2000	$6/MMBtu; 1975 dollars	0.99	0.8	0.5	0.67
		$2	0.01	0.2	0.5	0.33
	(With 4% annual OPEC price rise, $6 AES is not competitive with imported oil until 2010.)					
3. 1990	3. Coal Constraints	40 quads in 2000; 60 quads asymptote	0.7	0.9	1.0	0.75
		none	0.3	0.1	0	0.25
	(13 quads consumed in 1970 and 1975.)					
4. 1990	4. Limit on LWR capacity in 2000	400 GW	1.0	0.5	0.2	0.2
		300 GW	0	0.4	0.7	0.6
		150 GW	0	0.1	0.1	0.2
5. (i) 1900, or (ii) 2000, depending on timing of ACR Demo program	5. Annual uranium requirements for ACR (short tons of U_3O_8/GW-year, 65% capacity factor). (With 1% annual evolutionary improvements, open cycle LWR uranium requirements may be reduced to 122 ST/year.)	A. 96 ST/year (e.g., HWR, uranium, once-through)	1.0 0.8	0.99 0.9	0.7 0.5	0.9 0.1
		B. 30 ST/year (e.g., HWR, uranium-thorium, recycle)	0 0.2	0.01 0.1	0.3 0.5	0.1 0.9
6. 2000	6. Uranium resources (cumulative million short tons of U_3O_8)	Selling price ($/lb) $50 $150 $1000				
		Low 1.5 3.0 ∞	0.75	0.05	0.2	0.2
		Medium 3.0 6.0 ∞	0.25	0.9	0.6	0.6
		High 10.0 ∞ ∞	0	0.05	0.2	0.2
7. 2010	7. Costs of early advanced electric (fusion, solar, coal-solar, hybrid); option of 6 GW in 2010	2.5 x LWR costs	0.99	0.95	0.3	0.67
		1.25 x LWR costs	0.01	0.05	0.7	0.33
8. 1990	8. U. S. willingness to pay for avoiding additional international proliferation/diversion impacts	0.1% Fraction of 21st century GNP worth giving up to avoid plutonium fuel cycles in U. S., but continuing with once-through LWR	0	0.3	1.0	0
		0%	0.5	0.2	0	1.0
		0.1%	0.3	1.0	0	0
		x%	0.2 X = 0.1%	0.4 X = 0.5%	0	0
	(1% of today's GNP is $20 billions per year.)					
9. 2010	9. Probability that ACR system is significantly more proliferation/diversion (P/D) resistant than plutonium fuel cycles = y = N.B. Let x be the P/D costs (if any) of plutonium fuel cycles. Then (1 − y) x is the expected P/D costs of ACR system.		0.1	0.2	0.5	0

Table 15-2. continued

5	6	7	8	9	10	11	12	13	14	15	16
0.5	0.4	0.3	0.65	0.8	0.4	0.8	0.9	0.7	0.4	0.9	0.1
0.5	0.6	0.7	0.35	0.2	0.6	0.2	0.1	0.3	0.6	0.1	0.9
0.33	0.7	0.7	0.5	0.99	0.7	0.9	0.99	0.8	0.3	0.9	0.5
0.67	0.3	0.3	0.5	0.01	0.3	0.1	0.01	0.2	0.7	0.1	0.5
1.0	0.8	0.8	0.8	0.9	0.9	0.9	0.99	0.9	0.9	1.0	1.0
0	0.2	0.2	0.2	0.1	0.1	0.1	0.01	0.1	0.1	0	0
0.01	0.3	0.1	0.4	0.3	0.5	0.6	0.6	0.8	0.5	0.7	0
0.89	0.5	0.8	0.5	0.6	0.4	0.4	0.4	0.2	0.5	0.3	0.5
0.1	0.2	0.1	0.1	0.1	0.1	0	0	0	0	0	0.5
0.9 0.5	0.5 0.5	0.5 0.4	0.9 0.5	0.5 0.2	0.6 0.4	1.0 0.9	0.99 0.9	0.9 0.7	0.8 0.7	0.99 0.9	0.5 0.5
0.1 0.5	0.5 0.5	0.5 0.6	0.1 0.6	0.5 0.8	0.4 0.6	0 0.1	0.01 0.1	0.1 0.3	0.2 0.3	0.01 0.1	0.5 0.5
0.1	0.1	0.1	0.15	0.2	0.1	0.4	0.3	0.4	0.09	0.5	0
0.5	0.5	0.6	0.7	0.6	0.3	0.5	0.5	0.4	0.75	0.3	0.5
0.3	0.4	0.3	0.15	0.2	0.6	0.1	0.2	0.2	0.2	0.2	0.5
0.5	0.5	0.9	0.8	0.99	0.3	1.0	0.99	0.9	0.4	0.99	0.5
0.5	0.5	0.1	0.2	0.01	0.7	0	0.01	0.1	0.6	0.01	0.5
0.5	0.15	0.75	0.1	0	1.0	0.1	0	0.4	0.5	0	0.5
0	0.6	0	0.8	0.5	0	0.7	0.5	0.3	0	0.2	0
0	0.1	0.25	0.1	0	0	0.2	0	0.3	0	0.3	0
0.5	0.15	0	0	0.5	0	0	0.5	0	0.5	0.5	0.5
X = 0.5%	X = 3%			X = -0.1%			X = 0.1%		X = 0.5%_	X = -2%	X = 0.1%
0.8	0.2	0.5	0.5	1.0	0.7	0	0	0	1.0	0	1.0

To represent these differing viewpoints, the sixteen participants in our poll were asked to assign *probabilities* not only to the seven key technoeconomic factors, but also to the geopolitical uncertainties represented by the eighth and ninth items in Table 15–2. Consider respondent 6, who perhaps represented the widest range of viewpoints on the plutonium issue. That person assigned a 15 percent probability to the view that the United States would be willing to give up 1 percent of its GNP to avoid the additional international proliferation/diversion impacts. He or she assigned a 60 percent probability to the position that the United States would be willing to give up none of its GNP for this purpose; a 10 percent probability that the United States would perceive this as reducing international pressures on energy resources, hence counting this as a benefit (a negative cost) equivalent to 1 percent of GNP; and a 15 percent probability that the United States would be willing to give up 3 percent of the GNP to avoid plutonium fuel cycles.

Question 9 was intended to elicit views on the relative merits of the ACR versus plutonium fuel cycles with respect to proliferation/diversion resistance. Some respondents perceived significant advantages of the ACR through isotopic denaturing and a reduction in the plutonium content of spent fuel rods. Others perceived little or no advantage, perhaps because of the continuous on-line refueling feature of the Canadian heavy water reactor and also the possibility of diversion of heavy water for the production of plutonium in clandestine reactors. Again consider respondent 6 who assigned only 20 percent to the probability that an ACR would be significantly more proliferation/diversion resistant than plutonium fuel cycles.

DECISION ANALYSIS RESULTS

Before presenting the decision analysis results from these probability assessments, several technical points should be mentioned.

All uncertainties are represented as discrete rather than continuous probability distributions. To reduce the cost of computations, we adopted dichotomies or trichotomies, even though it might have been preferable to allow more choices to the respondents. This was particularly unfortunate in the case of question 3, dealing with the constraints on the use of coal.

As a further means of reducing computation costs, there are a series of outcomes labeled with asterisks in Table 15–2. These are events in which both the FBR and ACR would yield little or no gross economic benefits. For simplicity, we have assumed that if any *one* of these events were to occur, the gross economic benefits would be zero. This leads to a large number of states-of-the-world in which we may have slightly *under*stated the benefits of the FBR and ACR technologies.

Each of the entries shown in the leftmost column of Table 15–2 represents our best guess as to the date of resolution of these various uncertainties. Except for question 5 (dealing with the ACR's technical characteristics), we did not allow for variations in the date itself.

On the magnitude of uranium resources (uncertainty 6), we performed a Bayesian assessment on the outcome of the National Uranium Resource Evaluation program.

It is supposed that U.S. policymakers will consistently follow a "rational" course of action that maximizes the long-run expected net benefits to the public at large. It is quite possible that political leaders are motivated by other factors—for example, the desire to be reelected or the desire to avoid divisive controversies over environmental impacts.

It is exceedingly difficult to phrase a social issues questionnaire so as to avoid cuing the respondents. We have done our best, but are well aware that this questionnaire fell short of complete objectivity.

Subject to all the foregoing caveats, the numerical results of this decision analysis are presented in Table 15–3. No special significance should be attached to the implied number of votes for or against the FBR demonstration program. The number of votes could easily have been changed by sampling from a different population. Note, however, that none of these probability assessments imply support for an ACR demo program, even though several of the respondents had been on record in support of the ACR as against the FBR.

Perhaps the most striking result in Table 15–3 is that there is very little difference in the overall outcome whether or not one allows for the international proliferation/diversion impacts of plutonium fuel cycles. To understand why this occurs, it is essential to examine the correlations between the probability responses of each individual. From Figure 15–5, note that there is a high correlation between those who believe in a high elasticity of demand substitution (hence a low energy growth rate) and those who believe that the United States will be willing to pay a sizable fraction of its GNP to avoid the proliferation/diversion impacts of plutonium fuel cycles. (The $\overline{R}^2 = 0.60$ for all sixteen respondents and 0.62 for the eleven participants who came from universities and government agencies.) The correlation is somewhat lower but nonetheless significantly positive ($\overline{R}^2 = 0.35$) for uranium resource estimates (Figure 15–6).

Our overall conclusions are summarized by Figure 15–7. This chart indicates that those who assign a low proliferation and diver-

Table 15–3. Expected Money Value of the Economic Benefits *($ billions, discounted to 1975 at 5 percent per year)*

Excluding International Proliferation and Diversion Costs (or Benefits) of Plutonium in U.S.

	RESPONDENT															
	1	*2*	*3*	*4*	*5*	*6*	*7*	*8*	*9*	*10*	*11*	*12*	*13*	*14*	*15*	*16*
Today's Decision =																
FBR & ACR Demo	97.8	33.0	-2.9	10.2	0.5	0.0	3.8	10.7	49.7	-2.5	71.8	81.3	45.1	-2.2	92.6	-4.7
FBR Demo	(99.1)	(34.3)	-1.6	(11.5)	(1.8)	1.3	(5.1)	(12.0)	(51.0)	-1.2	(73.1)	(82.6)	(46.4)	-0.9	(93.9)	-3.4
ACR Demo	50.2	27.8	-1.9	7.0	0.3	0.3	3.3	8.2	34.0	-1.6	42.6	50.6	25.8	-1.0	49.5	-2.9
Wait	51.5	29.1	-0.6	8.3	1.6	(1.6)	4.6	9.5	35.3	-0.3	43.9	51.9	27.1	(0.3)	50.8	-1.6
Stop	0.0	0.0	(0.0)	0.0	0.0	0.0	0.0	0.0	0.0	(0.0)	0.0	0.0	0.0	0.0	0.0	(0.0)

Including International Proliferation and Diversion Costs (or Benefits) of Plutonium in U.S.

	RESPONDENT															
	1	*2*	*3*	*4*	*5*	*6*	*7*	*8*	*9*	*10*	*11*	*12*	*13*	*14*	*15*	*16*
Today's Decision =																
FBR & ACR Demo	141.2	17.3	-5.1	10.2	-4.8	-0.2	3.1	13.2	56.1	-5.1	93.4	72.6	52.3	-4.4	302.0	-4.9
FBR Demo	(142.5)	(18.6)	-3.8	(11.5)	-3.6	1.1	(4.4)	(14.5)	(57.4)	-3.8	(94.7)	(73.9)	(53.6)	-3.4	(303.3)	-3.6
ACR Demo	79.1	13.4	-3.3	7.0	-3.2	0.1	2.3	9.7	38.2	-3.3	57.5	44.7	32.3	-2.6	188.8	-3.1
Wait	80.4	14.7	-2.0	8.3	-2.0	(1.4)	3.6	11.0	39.5	-2.0	58.8	46.0	33.6	-1.6	190.1	-1.8
Stop	0.0	0.0	(0.0)	0.0	(0.0)	0.0	0.0	0.0	0.0	(0.0)	0.0	0.0	0.0	(0.0)	0.0	(0.0)

Figure 15—5. Results of Poll on Probability Estimates (16 Respondents):
Elasticity of Demand Substitution

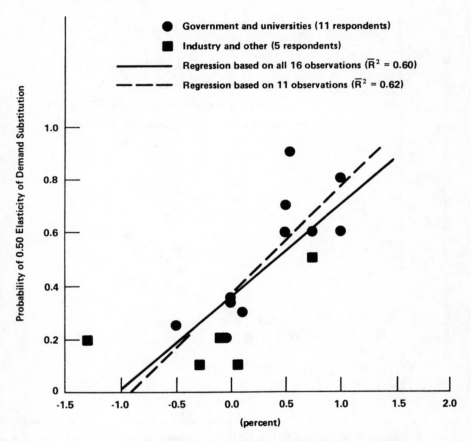

U. S. Willingness to Pay for Avoiding Additional International
Proliferation/Diversion Impacts (expected percentage of 21st century GNP)

Figure 15–6.   Results of Poll on Probability Estimates (16 Respondents): Uranium Resources

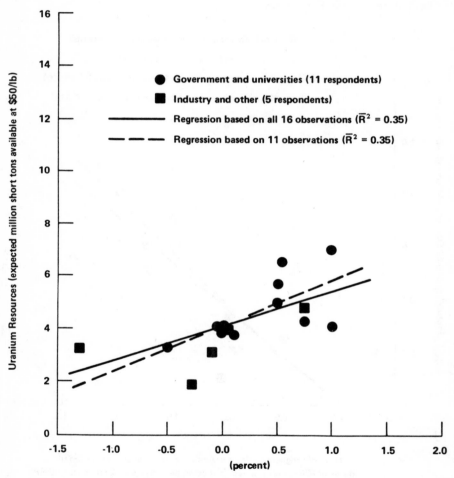

U. S. Willingness to Pay for Avoiding Additional International
Proliferation/Diversion Impacts (expected percentage of 21st century GNP)

Figure 15–7. Results of Poll on Probability Estimates (16 Respondents): Net Economic Benefits of FBR Demo Decision

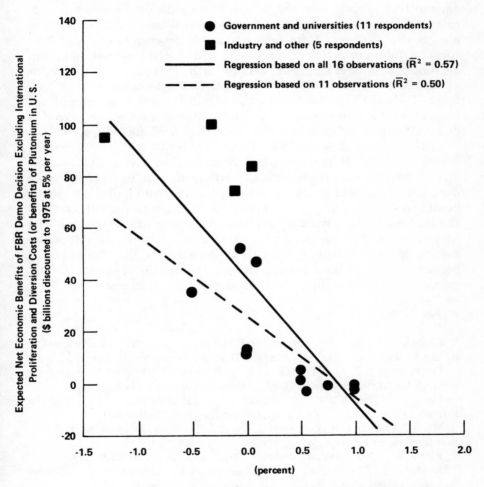

U. S. Willingness to Pay for Avoiding Additional International
Proliferation/Diversion Impacts (expected percentage of 21st century GNP)

sion cost also tend to ascribe high *economic* benefits to an FBR demonstration program—and conversely. In other words, this is a highly polarized debate. If one is for or against plutonium fuel cycles, one tends to adopt arguments on uranium and on demand growth that also support one or the other position. In debates such as this one, we do not believe that unbiased scientific judgments are likely to emerge from expert working groups. When individual topics are delegated to expert committees, the same general issues tend to recur in a somewhat different form (and perhaps in an even more frustrating form) within a uranium resources group, a demand estimation group, an economic modeling group, and so forth. There is a moral here for efforts like the Committee on Nuclear and Alternative Energy Systems (CONAES) of the National Research Council. One should not have expected a committee as polarized as this one to have arrived at coherent expert judgments within individual working groups.

This moral applies not only to CONAES but also to the current efforts of INFCE (International Nuclear Fuel Cycle Evaluation). With over fifty sovereign nations participating in this program, we doubt that one will obtain unbiased evidence from individual expert committees. Consciously or subconsciously, each group will confront the larger question. We are not altogether pessimistic about this kind of process, for it can provide a valuable educational experience to individual participants. From this type of effort, one can hope for greater understanding of alternative viewpoints, but it is unduly optimistic to expect a group consensus to emerge in this way.

REFERENCES

Cochran, T. 1974. *The Liquid Metal Fast Breeder Reactor: An Environmental and Economic Critique.* Washington, D.C.: Resources for the Future.

Feiveson, H.; F. von Hippel; and R. Williams. 1978. "An Evolutionary Strategy for Nuclear Power." Princeton University. Unpublished.

Keeny, S.M., et al. 1977. *Nuclear Power Issues and Choices.* Report of the Nuclear Energy Policy Study Group. Cambridge, Mass.: Ballinger.

Manne, A.S., and R.G. Richels. 1978. "A Decision Analysis of the U.S. Breeder Reactor Program." *Energy* 3, no. 6: 747–68.

Raiffa, H. 1968. *Decision Analysis.* Reading, Mass.: Addison-Wesley.

Rockefeller Foundation. 1978. "International Cooperation on Breeder Reactors." May.

Rose, David J., and Richard K. Lester. 1978. "Nuclear Power, Nuclear Weapons and International Stability." *Scientific American*, April, pp. 45–57.

✳ *Chapter 16*

The Political Point of View

Leon N. Lindberg

Presumably, introducing the political point of view implies introducing conflict into our discussions, and I am going to do my best to live up to that premise. My remarks will emphasize areas of disagreement with the papers in this session, as well as with aspects of the symposium as a whole. It is important to begin by saying something generally about policy analysis and the study of policymaking because what we are trying to achieve in this symposium is, to quote the letter of invitation, "a better understanding of the methodological and philosophical conflicts with which energy policymaking is beset." This implies that we begin by elucidating some of the ways in which different disciplines approach the study of policy questions.

Policy analysis has a number of elements of which four identified by Duncan MacCrae (1976) are significant for our purposes. The first is the procedure used to define a problem. How do various problems as defined by *actors* get to be perceived as problems by the analyst? It is a matter of art, not methodology, a matter of preference, values and ideology, and we have been befuddled by it to some extent throughout the symposium. The second element of policy analysis is the methodology for formulating clearly the values inherent in criteria or prescriptive methodologies—in other words, the use of clearly defined ethical paradigms. Different disciplines embody different ethical paradigms, a point on which I will comment further. The third element is the method for comparing alternative policies as predicted models of causation linking policies, consequences, and evaluative criteria. The fourth element is concerned with considerations of political feasibility, prospects for enactment and implementation, and the issue of how to improve policymaking institutions and procedures.

Broadly speaking, there are two sometimes competing approaches (disciplinary-professional clusters) in policy analysis. I will be very brutal and general and call them the managerial and behavioral sciences following the terminology of Y. Dror (1971). These approaches differ in very marked ways in their procedures with respect to the four elements I have enumerated.

The managerial or management sciences, by which I mean microeconomics, operations research, decision theory, management cybernetics, information theory, and systems analysis, is one body. These disciplines share a number of philosophical and methodological assumptions: a conception of rationality, a certain deductive style of analysis, a certain view of the significance of human decisionmaking, theories of utility, efficiency as an overriding criterion, and a commitment to quantification.

The second broad cluster of professions, the behavioral sciences, includes political science, sociology, anthropology, psychology, and *some* economics. (I put Dr. Roberts' work in this category, but I do not think of most economists as being very strong on the behavioral side.) The characteristics of the behavioral sciences are an inductive methodology and an emphasis on multiple values, especially humanistic ones, including the good life, social justice, social integration, public interest, and democratic participation in decisionmaking.

I think it is fair to say that microeconomic reasoning and the ethics that go along with the management sciences (cost/benefit analysis, the old and new welfare economics, and so on) seem to dominate policy analysis, at least in the United States, and certainly have done so in most of the presentations in this symposium. They dominate for a number of obvious reasons: the management sciences have a high degree of formalization, a specific mathematical notation, and a consistent prescriptive theory from which you can deduce a number of specific instances. But there are dangers in this domination of the agenda, and I think the preceding papers in this session illustrated this, especially the paper by Dr. Manne and Dr. Richels.

To quote Duncan MacCrae, who has written about this dominance of the economic paradigm: "some of its deficiencies are also those of American policy formation. . . . although we cannot easily replace the scientific paradigm of economics there is reason to believe that its ethical paradigms can and will be superseded . . . " (1976: 272). Indeed, Dr. Roberts' paper contains arguments along those lines.

What is needed is for the behavioral sciences to put their own shop in order, to perfect their own scientific and ethical paradigms, so that we can advance what MacCrae terms "dialogue among rival ethical hypotheses." If we do not do this, we allow a transfer of economic ethics to the realm of politics.

Economics has to do with the satisfaction of existing individual preferences and systems of exchange created to satisfy these preferences. Political science and political sociology are concerned with institutions in which responsible citizens and public officials consider the general welfare: with social norms, the "social contract," and "rational ethics." I would like to juxtapose these different ethical approaches and their associated empirical approaches, with reference to the symposium as a whole, then add a few substantive points stimulated by the symposium's papers.

REFLECTIONS ON THE SYMPOSIUM
AS A WHOLE

It is unfortunate that the formal papers in the symposium are so fully dominated by a management sciences approach, unfortunate because there are a number of well-known inadequacies with regard to policy analysis that attach to the management sciences. It is unfair to put Marc Roberts in this category, but I will leave him there for the time being.

The management science approach tries to propose optimal policies without taking into account the institutional contexts of problems, policymaking, and implementation processes. It is unable to handle political needs such as consensus-maintenance and coalition-building. There is inability to deal with "irrational" phenomena, ideologies, high-risk commitment, self-sacrifice, altruism, and reciprocity. Basic value issues that cannot satisfy assumptions of transitivity or a common metric fall outside its scope. There is a tendency to accept available alternatives rather than to invent new ones, and this implies a certain acceptance of the status quo. Predictability is required even in that sensitive analysis of uncertainty presented by Dr. Manne and Dr. Richels in Chapter 15. Primary uncertainty, where both the range of alternatives and the outcomes of those alternatives are unknown, is difficult to deal with.

Finally, the management science approach depends on the quantification of the main relevant variables and, therefore, is often unable to deal satisfactorily with complex social issues. And what Dror (1971) calls basic "megapolicy" choices with regard to risk and time tend to be taken for granted: maxi-min or mini-max as far as risk, discounting the future as far as time. And perhaps, most important of all for our purposes here, the problems of improving institutional capabilities and of increasing the scope of policy choice are neglected.

I am not making a blanket attack on decision theory or rational analysis by listing these inadequacies. But I think there is a very serious problem with these approaches when we are dealing with an area in which the main policies cannot be taken for granted, and the spontaneous adjustment of policy to changing needs cannot be relied upon. This is precisely the case with energy and, in particular, nuclear energy.

When the main policies themselves are in need of formulation or reform, improved managerial decisions can be useless and counterproductive. Better logistics for wrong wars, improved programming for projects when the structure of the area and of policy in general should be changed, tend to divert attention from the real problems. Another way of putting this is to characterize the agenda of this session, and of the symposium as a whole, as rather technocratic. Maybe there is a hidden agenda that underlies our deliberations. Let me give you a few examples that trouble me as a political scientist.

We are seeking a better understanding of "philosophical and methodological conflicts," to quote the words of the conference organizers. Does this tend to assume away issues of political conflicts, of power or domination-making process susceptible to constructive analysis? I am not sure what destructive analysis is, but maybe that is what I am engaged in at the moment. We are called upon to focus upon differences among disciplines in defining terms and concepts, divergences in interpretations and emphasis that cause confusion and mistrust of experts. Dr. Laney, in his remarks yesterday, accused the Daly and Malès papers of creating confusion. I would suggest that one should not mistake value conflict for confusion. The "confused public" is an interesting term; it implies, "Here is this nice technology about which the public is confused." The problem is to reduce this level of confusion.

I am reminded of a conference in Paris several years ago, on the psychosociology of nuclear power, which was really an effort to try to figure out what was wrong with people who had all of these irrational fears. I am afraid that there is some kind of agenda like that behind our deliberations as well. When you identify and emphasize mistrust of experts, you may badly misidentify the real underlying problem. The public opinion polls that I have seen show a mistrust not of science, not of experts, but of the ends to which science and expert advice are put, and of the political power of the politicans who manipulate these things.

Technocratic analysis performs a number of societal roles. It performs the ones emphasized on the agenda here: reducing uncertainty for policymakers, and improving the quality and quantity of information. It also legitimizes, ratifies, certifies, and rationalizes decisions already made, precisely because it considers only those options already defined. Technocratic analysis diffuses political conflict by focusing analysis on technical matters. It also stresses a "managerial rationality" (even in Dr. Roberts' paper), which is often a means to centralize power in executive managers. PPBS (planning-programming-budgeting systems), the new economics, the social indicator movement, all fall under that charge to some extent.

Thus, technocratic approaches have a number of costs. They minimize distributional issues; they treat politics as a matter of technique. The notion is that as the sphere of knowledge grows, the sphere of politics will diminish. They assume that better information can resolve value and political conflicts; they neglect questions of domination, power, and political struggle.

All of this is confirmed for me by the absence from our agenda of what, in my mind, ought to be our central preoccupation, namely, the abominable record in the United States and elsewhere with regard to nuclear energy decisionmaking, referred to by Bupp and Derian in their book *Light Water* as the nuclear debacle. They wrote:

> the way that the innovation process for Light Water reactors was managed by business and government in the United States and Western Europe contributed to the identification of nuclear power technology with something that many citizens in these countries dislike and distrust about their societies. . . . it is this dislike and distrust which is the driving force behind the nuclear safety controversy and the principal cause for the dissolution of the nuclear dream. (1978:11)

These last words, I think, are overstated. I do not think it is dissolving. I have seen all kinds of evidence in the past few days that it is not dissolving. To quote further:

> Contrary to the assumptions of government and business leaders, this opposition is not likely to disappear when the facts are established and commonly understood (p. 130). . . . For nearly a quarter of a century, the theology of nuclear power—unchallenged and unchallengeable—was accepted by a variety of diverse interests to advance a variety of diverse causes. Rarely did those who seized on nuclear power as a means to their own ends know its actual economic and technical status. Instead, the information available to them was part of a catechism whose basic function was to answer the questions of the infidels and sustain the faith of the converted. The result, a circular flow of self-congratulatory claims, preserved

the discrepancy between promise and performance. . . . Systematic confusion of expectation with fact, of hope with reality, has been the most characteristic feature of the entire thirty-year effort to develop nuclear power. It was the unintentional result of consciously designed institutional relationships among American and Western European scientists, public administrators, politicians and business executives. The economic "analyses" which controlled discussion during the critical early years of light water commercial sales had nothing to do with the detached confrontation of propositions with evidence. . . . The public agencies with putative responsibility for facing the facts had neither the means nor the motivation to respond critically to the nuclear industry's propaganda; they could only sanctify it. This they did with notable eagerness. By pursuing institutional interests only distantly related to those of the public at large, the Atomic Energy Commission and the Joint Committee on Atomic Energy became soap-boxes for light water promotional literature. . . . (p. 188–89)

Bupp and Derian are not opponents of nuclear power, but are favorable on technical grounds. Their book is about the *political* abuse of the technology. It is surely this record, and those characteristics of nuclear technology that seem to require centralization, secrecy, information control, and quasimilitary security systems, that convince so many that to accept nuclear power is to engage in a Faustian bargain that leaves control of nuclear assessment and decisions in the hands of those people with the most to gain from nuclear exploration. This is, of course, what political arrangements have done in the past. I am not sure how convinced we can be that the AEC, reborn in ERDA, reborn in DOE, escapes these criticisms today.

This raises a question that is not on our agenda: the acceptability of the decisionmaking process itself, from the point of view of democratic norms as a policy criterion. For me, the most problematic aspects of the energy decisionmaking process are the characteristics of the system that have constrained and continued to constrain public choice, and the existence of a de facto private-public energy plan, based on coal and nuclear power. I think that the Shepsle–Roberts–Lave analyses of yesterday and today really oversimplify. There *is* an energy policy in the United States, but it is not a publicly debated and arrived-at energy policy. I am very concerned with the whole question of the political stalemate on *public* energy policy. That such a stalemate exists is certainly clear from the Shepsle and Roberts presentations, but I think that the reasons they offer for it are wrong, and I will come back to it later.

Most important as a problem for me is the question how to create option space—how to create a feasible choice for the kinds of alternatives that Herman Daly was talking to us about, for a conservation

or a solar energy path. Perhaps the only way you can make plutonium acceptable is by convincing people that they have or have not
had some kind of real choice in the matter.

SUBSTANTIVE POINTS RAISED
BY THE SYMPOSIUM PAPERS

I would like to say something about the implied causal model of the
policy process that Dr. Manne's and Dr. Richels' very interesting
analysis of "decision trees" raises. I would also like to comment on
the model of the economy that underlies that paper and Dr. Roberts'
paper. This will lead me to some words about politics, rationality,
and the theory of the state, because these issues have been raised.
Finally, I want to say a bit more about what I think our agenda
ought to be.

The assertion was made that decision analysis helps design a flexible strategy for evaluating options, taking advantage of the sequential nature of decision and information flows. This assertion implies a
particular causal model of the policy process. There are several things
about this that I find objectionable. First, decision analysis evaluates
options in terms of economic risks only, upside and downside risks.
Other risks are not considered. Let me mention a few. Opportunity
costs of alternative energy sources are not explored, nor is the political commitment risk, the fact that as you move resources into a
particular technological area, groups form around it and make it difficult to change options at some future stage. Nor are the force-out
risks considered—that is, where one technology forces out another
by virtue of capital costs, research and development allocations, and
so forth. Finally, the costs of energy stalemate and energy polariza-
tions are virtually ignored, but these are likely to lead to real crises
in the 1990s, when I suspect we will have very few real choices available to us.

Now, as a model of the policy implementation process, decision
analysis seems to be deficient. First, it grossly underestimates the difficulties of keeping options open. There is no understanding of the
constant process of commitment due to the momentum of past policy, due to the structure of the energy system and the way in which
that energy system evolves, for example, in the direction of greater
electrification, and due to the structure of organizational rules that
tie together the different actors and institutions in an energy system.
There is a natural propensity of actors to seek predictability, and this
has to be factored in, if we are not going to stop at existing status
quo options. This approach neglects the fact that there is an estab-

lished energy policy trajectory that actively disadvantages certain kinds of options. The recent Battelle study of incentives for stimulating production of existing forms of energy (Battelle Northwest Pacific Laboratories 1978) seems to demonstrate this point very well.

There is an implicit bias in public and private energy policy towards supply-side rather than demand-side solutions, towards hard rather than soft paths. I think that this bias is much more pervasive and much more difficult to remove than is recognized by decision analysis. Incremental, sequential steps, such as are described in these decision trees, do affect power relations, do intensify disparities of power. This is clearly seen in the early history of nuclear decisionmaking in the United States, as described by Bupp and Derian (1978).

I would like to say something about the myths and stylized facts that underlie the model of the economy that one finds in the Manne and Richels paper. I do not think that this is a matter of artifice; I think that, to some extent, these myths or stylized facts are required by this kind of analysis. The ETA—MACRO model enshrines a free market model with exogenous politicians; a view of the political and economic adjustment process that is, in my view, politically naive— an insufficient analysis of the nature of the actors in the political economy, of the structure of institutions in that political economy, and of the incentives to which they respond. Economic efficiency is an explicit criterion and it becomes the de facto criterion because methods do not exist for trading it off with other values. Other criteria then get to be symbolically recognized rather than integrated in any real fashion into the analysis. Elsewhere, Dr. Roberts has argued that "energy is a commodity whose ultimate allocation can be left mainly to market decisions, in other words that most decisions of quantity and kinds of energy desired can be left to the market place." It seems to me that this view totally ignores the structure of power in the economy and in the political economy, and needs to be juxtaposed to experience with nuclear decisionmaking.

Finally, there is a concept of energy demand as an exogenously given variable that is an underlying assumption in Dr. Manne's and Dr. Richels' work, that drives the whole model, and that entirely determines the cost/benefit results—the upside and downside calculations. This is rather arbitrary, and there are many alternative demand parameters that one might have chosen. I refer to the CONAES demand panel as just one example.

I would next like to say something about politics and the theory of the state. Economic analysis very typically relegates the following kinds of roles to government: assure that prices reflect all costs, resolve value conflicts by providing a social welfare function ("society" has to decide), adjust incentives of private actors when these produce market failures, and so on. When politics, or the policy process, or politicians do not do these things, the standard explanation is along the lines of Dr. Roberts' paper and Dr. Shepsle's comments, both of which are very good examples of the genre. I like to call this a standard liberal pluralist analysis. It finds the cause of policymaking failure in group struggles for power, competitive bidding for votes, institutional decentralization, a system that protects minority views, and so on. I tend to view this kind of system as a pathology, as I think that Dr. Roberts does; yet Dr. Shepsle made it sound like a positive norm.

It seems to me that the standard liberal pluralist analysis, although not incorrect, substantially neglects the elite and the class dimension in power and policymaking, and the existence of what I call hierarchies of control. Whenever I hear the comment, "The American people are trying to make up their minds about energy," that seems to me a gross simplification of the way in which power is, in fact, distributed in this country. When I hear it said, as Dr. Lave did yesterday, that democracy is characterized by inertia, and the problem is one of overcoming the inertia of democracy, I would like to put the term "oligarchy" in place of democracy. I think it might come closer to describing the reality of energy policymaking in the United States, as I see it.

I think this pluralist analysis neglects the role of private power, what Lindblom (1977) has called the "privileged position of business" as the source of governmental pathologies. I do not deny the existence of the government pathologies that Dr. Roberts and Dr. Shepsle have chronicled, but I think their causation is much more complex and much more difficult to unearth.

I quote from William Ophuls: "Full disclosure of the kind of information needed to internalize the cost of production and make intelligent decisions on future developments is thus deeply threatening to the industrial order and to a political and economic system that has thrived on the invisibility of the hidden hand" (1977:178). This is not a Marxist speaking, but a rather traditional political philosopher. I try in my own work to suggest what these hierarchies of control are. I have studied the energy policymaking systems of seven different countries over the thirty-odd post World-War-II years, and I find a number of common characteristics. These include a closed circle of deciders; resistance to the representation of other actors; supply-oriented and technological-fix policy criteria; a resistance to alternative criteria and to public discussion of technological choices; sectoral and incremental decisionmaking leading to persistent suboptimization; no capacity to coordinate the system, no effort to seek new bases of higher-level value integration. Some of these you will be aware of from Dr. Roberts' paper.

I have a hierarchical structure of explanations. One of these is that interest group struggles and short-term electoral strategies preclude policy coordination and more than incremental change. Second, dominating elites and organizations in the energy supply sector are able to set limits on participation, restrict policy criteria, and defend their sectoral interests. At the third level of constraint, dynamically conservative bureaucracies act to retard information feedback and, therefore, change in effective loci of decision or criteria. This reinforces sectoral decisionmaking, and the access of private interests. Fourth, the policy role of technological expertise and technocratic culture retards change in the locus of decision and in policy criteria, and makes public participation in technological decisions more difficult. Fifth, dominant class interests in rapid growth and capital accumulation encourage resistance to changing deciders or decision criteria, and prevent the consideration of alternative development and growth strategies. This is nowhere more evident than in Eastern Europe or the Soviet Union. (My analysis is based on a study of Hungary, France, United States, Canada, Sweden, and India, and is not merely a description of capitalism. I am not flailing capitalism with this analysis.) Note that the very interesting reform suggestions that Dr. Roberts made at the end of his paper do very little to address these aspects of the problem of energy decisionmaking.

Let me conclude by reiterating my agenda for energy decisionmaking research and for reforming the decisionmaking process. First, devise fully collective and consensual approaches to policymaking. Second, develop new analytical and assessment technologies on the basis of an active dialogue among the competing ethical assumptions that are attached to disciplinary and professional clusters. Third, devise new ways and new political coalitions for using both markets and state power to control private sector power, private sector planning, and their administrative and scientific allies. Fourth, invent ways for assuring wider public participation and for avoiding technologies that intrinsically restrict it. Fifth, get away from paradigms such as those of some variants of decision theory that enshrine what I consider disfunctional myths and that ratify the status quo. Lastly, learn how to create option space—the possibility for choice—and understand more clearly how options may be inadvertently or deliberately foreclosed by the kinds of standard analyses that we have been offered in this symposium.

REFERENCES

Battelle Pacific Northwest Laboratories. 1978. *An Analysis of Federal Incentives Used to Stimulate Energy Production.* March.

Bupp, Irvin C., and Jean-Claude Derian. 1978. *Light Water.* New York: Basic Books.

Dror, Yehezkel. 1971. *Design for Policy Sciences.* New York: American Elsevier.

Lindberg, Leon N. 1977. *The Energy Syndrome.* Lexington, Mass.: Lexington Books.

Lindblom, C.E. 1977. *Politics and Markets.* New York: Basic Books.

MacCrae, Duncan. 1976. *The Social Function of Social Science.* New Haven, Conn.: Yale University Press.

Ophuls, William. 1977. *Ecology and The Politics of Scarcity.* San Francisco: Freeman.

✳ *Chapter 17*

The Psychological Point of View

M. Brewster Smith

During the heavy program of this symposium, we have been exposed to complex mapping schemes intended to help people come to grips with an enormously complex policy domain that cuts across very specialized specialties. We have also been offered several simplifying schemes: soft versus hard energy packages (Dr. Daly), Chicken Little versus Dr. Pangloss (Dr. Shepsle), we/they (Dr. Rochlin), and now managerial-technocratic versus behavioral-humanistic approaches (Dr. Lindberg). I was particularly helped by Dr. Lindberg's critical analysis of the conference agenda, which we have just heard. Dr. Lindberg knows the context of energy policy from close study, as I do not (in terms of the we/they dichotomy, I am one of *them*). So he has criticized the managerial-technocratic approach (as he observes, the major emphasis of the conference), in much more cogent terms than I could. But I can associate myself with his critique enthusiastically.

Our host and organizer, Dr. Sachs, expressed at the beginning his hope that we would be polite to each other and not let argument reach a "critical" level. Up until Dr. Lindberg's commentary, it seemed to me that his hopes for peacefulness had been too fully realized. We have not argued enough! The papers both by Dr. Manne and Dr. Richels and by Dr. Roberts give us a final chance—already responded to by Dr. Lindberg—to engage in the serious debate that the urgency and the complexity of our topic requires.

The rational, quantitative economic decision analysis presented by Dr. Manne and Richels is surely an example of the managerial-technocratic style at its most impressive. If one can avoid being bemused by its seeming precision and formalized authority, I am ready to grant that such a structure of decision trees should help to decompose issues in ways that could be useful in decisionmaking. But it is a big *if*: the hard, quantitative, rational style of this bravura performance is in such dramatic contrast with the extraordinary uncertainties and complexities that characterize the domain to which it is applied. So much of human value gets left out of the formal model; and the probabilities inserted in it by "Delphi" procedures are so subjective, so ungrounded and unstable.

As I understood in the preliminary paper that Dr. Manne circulated in advance of the symposium, his "Delphi" amounted simply to a questionnaire on a tiny "sample" of experts (without iterations to try to produce movement toward an informed consensus), which yielded probability estimates scattered all over the map. Surely such estimates, if used directly in the rational model, provide the shakiest basis for rational decision. In Dr. Manne's oral presentation, the divergent estimates of individual experts were used in what strikes me as a much more legitimate way: grinding them severally through the model to show, in terms of the model, the policy implications of different probability estimates. If participants are sufficiently sophisticated to grasp what the model assumes and to be alert to what it leaves out, I can believe that being confronted *both* with the diversity of estimates and with what the model says about the consequences that should follow from each could be an educational exercise that could raise the quality of policy discussion. Even so, it seems to me that the approach amounts more to a soft heuristic to clarify policy debate than to the hard, persuasive, scientific schema that its symbolism would make it appear.

I find the approach taken by Dr. Roberts' paper much more congenial. As a psychologist, I have been delighted to discover at this conference the existence of economists like Dr. Roberts and Dr. Daly, who use their tools of rational analysis with skill, but who retain what seems to me an invaluable sense of the limits of these tools, and a recognition of the actual social and human context in which difficult social decisions have to be made. Dr. Roberts' psychological assumptions fit well with mine.

Dr. Roberts' paper, as I read it, was more an excursion into political science than an exercise in abstract economics. So I defer to Dr. Lindberg for critique of his specific recommendations to improve governmental process. As a friendly observer of political science over the post World-War-II period, however, I like the way that Dr. Roberts combines a tough-minded perspective on the limitations of our political institutions and processes, with respect to producing rational policy (here he joins with Dr. Shepsle's remarks earlier in the symposium), with some probably less tough-minded suggestions as to what might be done to improve their functioning. If the American government is ill-designed to produce long-term rational policy, and it surely is, that should chasten the rational model builders. But it should not inhibit people with a vision from continually trying to make our system of social decision processes work better.

I warmed particularly to Dr. Roberts' advice to *us*: in favor of modesty, and of explicitness about our values. The managerial technocratic style is full of hubris. And we can discuss value disagreements usefully only when we are explicit about them. Little that is useful is likely to emerge when value arguments are disguised as arguments about facts. (For a thoughtful examination of these matters, and some proposals for constructive controversy about values, see MacCrae 1976.)

I worry somewhat about Dr. Roberts' use of Rousseau's old distinction between the "general will" and the "will of all." Much that is totalitarian and evil, in retrospect, has been done in the name of the "general will." It is admirable to try to get citizens to consider the public interest as well as private advantage, but it is dangerous to trust uncritically people who are sure that they know what the public interest is. We are all gifted rationalizers! Along with John Gardner's Common Cause, I would put more weight on procedural reforms that seem likely to give competing proposals a fairer hearing. I have similar doubts, by the way, about Dr. Daly's recourse yesterday to "objective values." Many of us may feel that we know them, but this feeling is all too subjective! In the sphere of values and the public interest, it is through a continuing social dialectic of ever-reformulated questions, proposals, and facts (and less rational components of rhetoric and politics) that we arrive at consensus or compromise, when we do so, or, which is just as important, at a transformation of the issues that divide us.

Let me recur to *them* (the nonexpert citizenry) and how *they* enter the process of social decision. Political scientists and sociologists are more sensitized than psychologists to the problem of how to *compose* (not just to aggregate) the different perspectives of citizens in the political process that deals with energy issues. But accurate information from surveys about the terms in which people are actually thinking about energy alternatives, including plutonium, is surely needed. The Battelle group in Seattle have recently integrated currently available survey data on energy alternatives and done exploratory work of their own on values involved in policy preferences regarding nuclear waste disposal (Melber et al., 1977; Maynard et al. 1976).

In a nutshell, it turns out that *they*—the general public—are now thinking along lines that do *not* pit one energy alternative against another. The basic polarity, in recent surveys, is rather between those who would cut back on energy demand, and those who favor economic growth and increase of energy supply on all fronts. The supporters of nuclear energy tend also to be the supporters of coal and solar alternatives. This line of cleavage sounds to me not too different from the choices offered in Dr. Daly's paper, or, if I understood them correctly, like that turned up by the "Delphi" among Dr. Manne's and Dr. Richels' experts. Maybe *they* are not so naive after all! Maybe we are in the early stages of a serious national debate on the values linked to economic growth—a good distance away from the special and crucial problems of plutonium, but perhaps just as fundamental for our future.

This leads me back to the general issue posed by many papers in the symposium, particularly by those of Dr. Manne and Dr. Richels and of Dr. Daly: "What is the actual and what is the appropriate role of 'rationality' in social choice?" Our answers do, indeed, involve a theory of human nature, explicit or implicit. To evaluate them responsibly would require a much more far-reaching discussion than is possible here.

A psychologist would have to insist that decisionmakers and people at large are only very partly rational. The questions that get posed for rational or less rational decisions, personally and socially, do not themselves arise from a process that can be rationally formulized. Our decisions, personally and politically, are not fully consistent with one another, and it is probably folly to pretend that they can be. Economic formulas, as an attempt to promote rational consistency—as a device of political rhetoric, if you will—constrict the value context of controversy that might otherwise be of human concern. (The introduction of discount rates imposes the marketplace framework on value issues concerning responsibility to our successors. We might resolve them quite differently if they were presented in different terms.)

Because of the fact that they center on questions of value, moreover, the issues that preoccupy politics do not fit the model of rational problemsolving in engineering technology (Sarason 1978). Often there is *no* technical solution to a persistent social problem; politics simply reworks the problem iteratively, in keeping with the shifting historical and evaluative context. The issues concerning nuclear energy and plutonium obviously have central technical ingredients. But it tends not to be so obvious to the technologists responsible for energy development, or to many social scientists who are bemused by the success story of the physical sciences, that they also have human ingredients that may defy purely technical solutions. How to bring technical skill from the social and behavioral sciences, as well as engineering competence, to bear in such distinctively political contexts poses dilemmas that we have by no means solved—an occasion for the modesty that Dr. Roberts calls for. At least, it urges extreme caution about irreversible commitments that bind the future by our own historically bounded decisions.

Finally, what some psychologists have come to call "hot decision-making" especially departs from rational formulae (see in particular Janis and Mann, 1977). Issues carrying such fearsome consequences as those clustering around plutonium are sure to evoke potentially self-defeating forms of irrationality well beyond people's characteristic noncompliance to the rational economic model—such pathologies as leaping at insufficiently examined alternatives, bolstering shaky choices to which one has become prematurely committed, and what Janis has called "group-think." There are measures that can be taken to increase the likelihood that our social decision process is rational, in this broader and consensually desirable sense. (Again, see Janis and Mann for some of the best-grounded present psychological wisdom.) I believe that this conference, with the discussion it has stimulated, is a step in the right direction.

REFERENCES

Janis, Irving, and Leon Mann. 1977. *Decision-Making: A Psychological Analysis of Conflict, Choice, and Commitment.* New York: Free Press.

MacCrae, Duncan, Jr. 1976. *The Social Function of Social Science.* New Haven: Yale University Press.

Maynard, W.S., et al. 1976. "Public Values Associated with Nuclear Waste Disposal." BNWL—1997. Seattle: Battelle Northwest Laboratories.

Melber, B.D., et al. 1977. "Nuclear Power and the Public: Analyses of Collected Survey Research." U.S. DOE Report PNL—2430. Seattle: Battelle Human Affairs Research Centers.

Sarason, S.B. 1978. "The Nature of Problem-Solving in Social Action." *American Psychologist* 33: 370—90.

✳ *Chapter 18*

The Anthropological Point of View

Stephen Beckerman

In preparing these remarks, I have tried to keep in mind the title of this symposium—"How Do We Decide?"—as well as the contents of the two very interesting and informative papers by Dr. Roberts and by Dr. Manne and Dr. Richels. I will first try to establish a sequence of related points:

1. There is no inevitability in the deployment of a particular technology simply because it exists.
2. However, once a major technological form is widely deployed, it does have major social, cultural, and political effects.
3. If I can demonstrate these two points to our satisfaction, it will follow that there is a "decision" to be made (a point that everyone seems to have assumed, so far in this session), and that this decision is one of widespread, legitimate general interest.
4. There will still remain the possibility, however, that although the sociocultural effects of the breeder reactor are of legitimate general concern, the criteria that must be used to make the decision to deploy or not to deploy are overwhelmingly technological or (what is a similarly monolithic position) overwhelmingly economic.
5. Therefore, I will spend a few moments arguing that from a *purely* technological point of view, there are no obstacles to the *immediate deployment* of the breeder reactor.
6. But from a purely economic point of view, there is no reason *ever* to deploy the breeder reactor.

7. It is from this perspective that I want to discuss papers by Dr. Roberts and by Dr. Manne and Dr. Richels.
8. I will integrate that discussion with an attempt at an appreciation of the magnitude and direction of the likely effects of the widespread deployment of a breeder energy system.
9. I will also try to integrate it with the underlying point that it is clearly and unarguably the case that we *cannot* have any kind of society we want in combination with any kind of energy system we end up with. The energy system sets strong, strict limits on the kind of society that is even possible, let alone likely.

We sometimes hear that naive statement that the breeder reactor is inevitable because all technological advances are more or less fully implemented as soon as they come along, because there is some sort of inherent drive to "efficiency" or "progress," whatever that means. Let me give you a few counterexamples—and let me add that these are just the first few examples that came to mind during this symposium. This is not an exhaustive list, by any means.

The Maya Indians, who built magnificiently elaborate pyramids, and had a calendar much more accurate than ours, had the wheel—and used it only on children's pull-toys (Porter-Weaver 1974).

The Chinese invented printing long before its invention in Europe, but even though they had invented movable type by the eleventh century, printing with movable type never replaced printing with large blocks, on which an entire page was carved or cast. Kroeber (1948: 492—95) has a fascinating discussion of this piece of technological history. The Chinese also invented gunpowder long before the Europeans, but did not produce guns to go with it until so late that they have apparently still not caught up with the West in firearms technology. Some of the relevant history is reviewed by Hall (1956: 377—79).

The crank and connecting rod, which surely has to rank on anyone's list of the dozen basic mechanical inventions (it is the basic means of transforming rotary to reciprocating linear motion, and vice versa), was invented in Europe at least as early as the fifteenth century, but according to Gille (1956: 654), "even in the seventeenth and eighteenth centuries, the crank and connecting rod were seldom combined." Mechanically inferior means of transforming rotary to linear motion dominated until the nineteenth century.

Turning to power technology, Forbes (1956) observes that the Romans had invented the vertical water wheel, the first invention that put in human hands more power from a single unit than could be obtained from a man or a draft animal, by the first century B.C.

However, the Roman world continued to rely on the horizontal water wheel, which had no greater power output per unit than a donkey or a horse, until "well into the third century A.D., and even then the vertical mill was restricted to a few centres only" (p. 601). Forbes also mentions that Vespasian (A.D. 69–79) is said to have refused to build a labor-saving water-driven hoist "lest the poor have no work" (1956: 601).

Finally, the Stirling engine, possibly the most thermodynamically sophisticated of the high-fuel-efficiency engines now being considered for automobiles, was invented over 150 years ago (for pumping out mines, I believe) and has hardly been anything but a laboratory toy until the last ten years. I think these examples constitute a plausible case for the noninevitability of technological deployment.

As far as the major social and political effects of technological deployment, we can take the turf-turning plow, which allowed the dense agricultural settlement of Northern Europe, shifted the populational and cultural center of gravity of the European peninsula, north from Rome towards France, Germany, and England, and was intimately involved with the rise and establishment of feudalism (cf. Singer 1957: 775). The stirrup is also generally held to be even more intimately involved with the rise of feudalism, and, in fact, it is widely believed that this invention led directly to the establishment of the Carolingian empire (White 1962).

Closer to home, and somewhat more specifically, we have only to look at a map of the United States to see what an enormous influence the railroads had on the placement of our towns and cities. And we have only to read biographies of such men as Andrew Mellon, J. Pierpont Morgan, Frank Norris, and Clarence Darrow, to see how vast was the influence of the railroads on every aspect of political and economic life—and how close that situation brought us to something very much like class warfare.

The extraordinary power of the railroads was broken in a number of ways, of which the most important were probably regulation and the rise of another technological innovation—the trackless, cheap, internal combustion-propelled vehicle, that is, the car, truck or bus. (It is symptomatic of the importance of this kind of vehicle to our society that we do not possess a single word to cover all its forms, just as the Eskimos do not have a single word for snow.)

There have been volumes written on the social influence of the automobile and its cousins. From the shape of our cities to the highway death toll, from residential segregation to the political role of the Teamsters' Union, from our patterns of internal migration and mobility to the economic and political strength of the UAW, from the influence of the "big three" automobile companies in domestic politics to the influence of the "big seven" oil companies in foreign policy, there is hardly a day in our lives in which we are not touched by the horseless carriage.

Now I submit that a technology like the breeder reactor, which is not inevitable, which bids fair to have an effect on our lives comparable to that of the automobile, and which will require both legislative change and the expenditure of large amounts of public moneys to deploy, is a legitimate and, indeed, necessary subject of deep and general public concern. However, there is still a question whether the criteria we must use, as members of the public, are monolithically technological or monolithically economic. The arguments usually made are either that we cannot have the breeder because it is inherently technologically unsafe, or that we must have the breeder because solar and biomass systems (the two main competitors) are too expensive. I submit that both these positions are specious.

If we take the question of technical safety, it is clear that reactor accidents are vanishingly unlikely to kill more than 50,000 people a year—and we already kill that many people a year with the car. I am not suggesting for a moment that an additional 50,000 deaths a year from breeder reactors would be acceptable to any of us or anyone else in this country. The point I am making is that the issue is not a technological one—it is a question of our political morality, our social psychology, and a number of other "soft" areas of our lives. These matters are not easily quantifiable, but they are very real.

It is also clear that we can reduce the risk of plutonium theft to a very, very low level by raising the level of armed security at reactor sites, reprocessing plant sites, waste disposal sites, and all the transportation routes between them, to the level now in force at our nuclear weapons bases. After all, no one (to the best of my knowledge) has even stolen one of our atomic bombs. But there is again a "soft" question here that governs all. Are we willing to live our lives in a country run like a nuclear weapons base?

On the other hand, let us consider the economic question. Take all the tax revenues that have been spent on nuclear development, both civilian and military, and express that money as a fraction of the GNP for the appropriate years. Add to that figure the tax revenues lost as a result of the oil depletion allowance, also expressed as a fraction of the GNP, for the years in which it has been in effect. Now add to that sum the tax revenues lost as a result of foreign tax credits to oil companies, again expressed as a fraction of the GNP for the appropriate years. Now, take that summed fraction of the *current* GNP and compute how many current dollars it is worth. It is very difficult to come up with hard figures, of course, since all these expenditures and losses are to some degree classified, confidential, or just plain never calculated. Still, it is my strong impression, that we are dealing with a sum of money so large that, if it were simply applied as a production-end subsidy to solar energy, it would make solar energy very cheap indeed. After all, the major cost of solar power is constructing the plants; the fuel is free.[a]

Having said that in strictly economic terms there would be plenty of money available to buy us the alternatives to the breeder reactor, let me add that, in more realistic, noneconomic terms there is no chance whatsoever of our spending all that money as a production-end subsidy to solar energy. That is as unlikely as our embracing an additional 50,000 deaths a year from radiation-induced cancers. The

[a]An hour or two after I spoke a gentleman from the audience very reasonably challenged me, over lunch, to produce at least some numbers to support my claim. I have found it every bit as difficult to encounter any numbers as I expected it to be, particularly in the short time allotted for the polishing of these presentations. Nevertheless, I will make an attempt at a very rough calculation. The sources I have had to use may cause a raised eyebrow or two. Lundberg (1968) following Stern (1964) figures that the oil depletion allowance plus intangible drilling deductions by the oil companies cost the federal treasury about $2 billion each year. I presume the base year for this figure is 1963, the year before Stern's book came out. If that is the case, the lost revenue would equal about one-third of 1 percent of the GNP of that year, which was $599 billion.

Sampson (1975, p. 245) asserts, without numbers, that the real tax advantage of the large, "Seven Sisters" oil companies comes these days, not from the depletion allowance, but from the foreign tax credits they deduct from their domestic taxes. Since these seven companies control about 70 percent of the world's oil production (excluding China and Eastern Europe) (Sampson 1975, p. 241), we are dealing with another significant sum. If (and this is just a guess) foreign tax credits are twice as significant as domestic depletion allowances, then we must add another two-thirds of a percentage point of the GNP, or rather 70 percent of that two-thirds, to account for the lost revenues. Let us take as a round figure that we have so far about three-quarters of 1 percent of the GNP.

On the nuclear end, Kaufman (1972, p. 38) reports that the AEC budget for 1969 (the GNP was then $947 billion) was $2.5 billion. That gives us about one-quarter of 1 percent of the GNP for that reference year. Finding the fraction of the military budget that goes for nuclear weapons and their support is very difficult. The total salary operations, maintenance, research, and procurement fraction of the military budget in 1969 was about $75 billion (Kaufman 1972: 41). If only 5 percent of that total went for nuclear weapons and their support maintenance, etc., we would get another one-third of 1 percent of the GNP of the reference year.

The grand total is about 1.3 percent of the GNP. Our current GNP is about $1.6 trillion. If we were willing to spend 1.3 percent of our GNP each year for the next twenty or twenty-five years, then we would have $400 billion 1978 to play with. (I would welcome a repetition of this exercise by someone with the time, patience, and expertise to go through military budgets, IRS documents, and congressional hearings transcripts, and come up with really accurate figures. Even so, I think the point is made. Even if my numbers are two or three times too large, we are still dealing in the order of magnitude of hundreds of billions of dollars.)

money may exist, but we have "chosen" to spend it on other things, on other kinds of subsidies, and that choice is, for all intents and purposes, irrevocable. And it is irrevocable for social and political reasons, as much as for economic factors.

Having come this far, and having convinced you, I hope, that neither technological nor economic considerations, alone or together, can or will be the whole basis on which we make our decision about the breeder reactor, I can finally turn to the two papers that I am here to review. Each is an excellent example of a kind of social systems' modeling. The first, by Dr. Roberts, is a qualitative, broad-ranging, but nevertheless somewhat formal and somewhat predictive model. The second, by Dr. Manne and Dr. Richels, is a quantitative, narrower, but highly formal and explicitly predictive model.

Despite my admiration for the ingenuity and complexity of both these models and for the amount of work that has gone into producing them, I am troubled by what they do not do. In fact, it is a common problem in the social sciences, as Bergmann (1975) remarks in another context, that models often, if not usually, tell us less than we already know. Our most effective political leaders, for instance, often profess outright contempt for formal models, on the plausible grounds that they can make better predictions, and operate more effectively, on their own "seat-of-the-pants" intuitions. It is a problem the social sciences will have to tackle head on, one day. How far can we go in modeling ourselves, when we, ourselves, are doing the modeling?

Now, turning to Dr. Roberts' paper, I am especially impressed by his use of Rousseau's distinction between the "will of all" (the summed expression of everyone's selfish desires) and the "general will" (the general consensus as to what a just society should be, putting aside selfish desires as much as possible). I will return to that distinction in a moment; right now, I want to make some less philosophical comments on his presentation.

Dr. Roberts' emphasis is on designing institutions to *do* something, which is all well and good, but as Dr. Rochlin remarked yesterday, "*Who* is going to design?" is left as an open question—open except for the portentous word "we." Again, what those institutions should do is simply assumed: we need an energy policy—a clear, consistent, uncontradictory energy policy. It is difficult for me to imagine a single, consistent energy policy that would serve the interests of all the "we's" in this country, so I conclude that some of the people are excluded from "we." I would like to know who they are.

I suspect that in the real world, we will get not a single energy policy (although it will be labeled as that), but a large set of policies that are often largely contradictory. We will pursue gasoline conservation, but that may make it difficult for people to get to work. We will augment our coal production, but that will conflict with our efforts to preserve the environment, and both of these efforts will have negative effects in stockpiling foreign oil, for instance, or building breeder reactors.

I doubt that we can have a consistent energy policy, any more than we can have a consistent tax code. Energy is as ubiquitous as money in our society, and different people's relationships to it are just as complex. No doubt, everyone would agree that present energy legislation needs reform; everyone agrees that tax legislation needs reform. But when we ask what the reforms should be, that splendid consensus ends in a big hurry. I think we had better resign ourselves to a hodgepodge, but a hodgepodge that will, I hope, at least not throw anyone into the energy equivalent of debtor's prison, or leave the country in the position of energy bankruptcy.

While Dr. Roberts was speaking, and particularly while he was lamenting the lack of effect of disinterested, nonpartisan opinion on the decisionmaking process, I began to mull over his distinction between the general will and the will of all. Particularly, I wondered what the will of all would be for this gathering. Suppose we took the most self-interested desires of each of us, with regard to energy, and summed them over the whole auditorium. What sorts of energy policy recommendations would we come up with?

Two obvious proposals spring to mind. First, it is clear that we would each profit from an increased role for scientific, academic, and technical expertise in the decisionmaking process. But to whom would we deliver our expertise? Clearly, it would be to our advantage to give it to some agency with the power to act on it decisively. Otherwise, we would be impotent. And, too, in a time of decreasing opportunities for academic and quasiacademic employment, it would be a good idea if this agency could employ some of us from time to time. And it would be most convenient if this employment could be effected without all the hassles, forms, delays, and picayune regulations of the current civil service. A strong top management structure in the federal executive would be an excellent candidate for the sort of agency we would like.

Of course, we are biased. Of course, Dr. Roberts' disinterested recommendation on the basis of the general will turns out to be the will of this assembly writ large. We are arguing about the conditions of our lives. How can we be anything but self-interested?

Dr. Manne and Dr. Richel have largely anticipated many of my comments on their paper, but I want to run over them briefly, be-

cause some of them apply to other papers given here, as well. Any
formal model in the social sciences requires a pickle-barrelful of simplifying assumptions just to make the analysis tractable. In Drs.
Manne and Richel's model in particular, a number of very real possibilities have been (I repeat, necessarily) assumed out of existence.

The effect of war on the projections is not considered, for instance, even though it has been a long time since we have seen a year
without a war somewhere in the world, and even though this country
has averaged one war every twenty years in this century, and the
rate, if anything, seems to be increasing. There is no consideration of
an interruption of fossil fuel supply, even though none of us here
would want to lay a hand over his heart and proclaim that another
Middle East boycott is impossible. (If there was ever a region ripe
for war, it is the Middle East in the last quarter of the twentieth century. What does it do to everyone's projections if somebody torches
the Saudi oil fields?) There is no consideration of the effect of a significant augmentation of fossil fuel supply, still a real possibility, in
view of what has happened on the North Slope, in Mexico, and in
the Ecuadorian Amazon over the last few years.

Then, there are the questions which Dr. Manne and Dr. Richels
ask their respondents. The questions themselves embody a view of
what is important and how things work, but the respondents are
not asked if they agree with that view. Again, as Dr. Manne and
Dr. Richels quite rightly note, the crucial dates for the resolution of
the uncertainties of their model are themselves only hopes. We really
do not know when these uncertainties will be resolved. Dr. Manne
and Dr. Richels are quite frank that the model tells us little and that
the answers they received from their respondents are expressions of
passion, special pleading, and ignorance, as much as informed and
objective judgment. Their conclusion is very much to the point. We
are passionate and self-interested, and even bigoted in this matter. I
would add that we should be. The issue is: In what kind of society
will we live our lives?" and we damned well should be passionate
about that.

One of the most thoughtful views of the kinds of societies implied
by nuclear power and by solar biomass power was published in 1978
in the *American Scientist* by Dr. Alvin Weinberg. I recommend it
to you. Dr. Weinberg sets up two polar sociocultural types, which he
calls the Solar Utopia and the Nuclear Utopia. The Solar Utopia, as I
read his exposition, is typified by a stout peasant yeoman, standing
with one hand on his horse-drawn plough, and the other hand on his
flat plate collector. Behind him stand a couple of apple-cheeked

children carrying handwoven baskets of organic foods, while his wife knits blankets under the apple tree.

The Nuclear Utopia, on the other hand, is typified by a hundred-odd radioactive sanctuaries of holy ground, consecrated in eternity to the nuclear enterprise. A purple-robed nuclear priesthood strolls serenely along stainless steel sidewalks, discussing Hegel and Schopenhauer, while a grim-faced nuclear papal guard keeps the angry masses from defiling the holy of holies.

These descriptions are exaggerations, caricatures. But they make a point we have to address. There is a choice to be made, and the choice is of something like this order of magnitude. Of course, we are passionate in the face of this kind of choice about how we are going to live our lives. We are as passionate as Jean Warren-Curry, who replied to Dr. Weinberg's article by writing (1978), "My question is, 'Can we do without Alvin Weinberg?' For myself and my daughter, the answer is a resounding YES." We are as passionate as Dr. Bernard Cohen, who proclaimed from the floor yesterday that the argument is over and that all we have to do is find some way to shut up guys like Robert Redford. We are as passionate as any human being must be, when confronted by a challenge to the quality and conditions of his own life.

I do not know whether the Romans passionately debated the virtues of the vertical water wheel, which they eschewed, or whether the denizens of the Dark Ages passionately debated the failings of the same source of power, which they embraced. I suspect that the engineers and artisans who stood to gain from this new source of power advanced forceful arguments in its favor, and the artisans and workmen who saw themselves endangered by it objected strenuously to its deployment. I further suspect that both sides saw right, justice, reason, and truth as their exclusive allies. I also suspect that in the clinches both sides fought dirty.

I have no predictions as to *what* we will decide with respect to the breeder reactor. I do have a prediction as to *how* we will decide. I submit that we will decide with appeals to our highest faculties of reason, and descents to our lowest capacities for dirty politics. We will decide on the basis of short-term interests in preserving our own wealth and power and long-term interests in preserving the human race. We will decide on the basis of personalities and lies, and prejudice and malice; and we will decide on the basis of altruism and facts, and reason and benevolence. We will, in short, be human, and I for one, will be very suspicious of anyone who claims superhuman disinterest for his own position on this question.

REFERENCES

Bergman, Frithjof. 1975. "On the Inadequacies of Functionalism and Structuralism." *Michigan Discussions in Anthropology* 1: 3–23.

Forbes, R.J. 1956. "Power." In Charles Singer et al., eds., *The History of Technology*, pp. 589–622. Oxford: Oxford University Press.

Gille, Bertrand. 1956. "Machines." In Charles Singer et al., eds., *The History of Technology*, pp. 629–58. Oxford: Oxford University Press.

Hall, A.R. 1956. "A Note on Military Pyrotechnics." In Charles Singer et al., eds., *The History of Technology*, pp. 374–82. Oxford: Oxford University Press.

Kaufman, Richard F. 1972. *The War Profiteers*. Garden City, N.Y.: Doubleday Anchor Books.

Kroeber, Alfred L. 1948. *Anthropology*, 2nd ed., New York: Harcourt, Brace.

Lundberg, Ferdinand. 1968. *The Rich and the Super-Rich*. New York: Bantam Books.

Porter-Weaver, Muriel. 1974. *The Aztec, The Maya, and Their Predecessors*. New York: Seminar Press.

Sampson, Anthony. 1975. *The Seven Sisters*. New York: Bantam Books.

Singer, Charles. 1956. "Epilogue: East and West in Retrospect." In Charles Singer et al., *The History of Technology*, pp. 753–76. Oxford: Oxford University Press.

Stern, Phillip M. 1964. *The Great Treasury Raid*. New York: Random House.

Warren-Curry, Jean. 1978. Letter to the Editor. *American Scientist* 66: 404.

Weinberg, Alvin M. 1978. "Reflections on the Energy Wars." *American Scientist* 66: 153–58.

White, Lynn T. 1962. *Medieval Technology and Social Change*. Oxford: Clarendon Press.

Discussion

Alan Manne: This morning has been a valuable occasion, one of those all too rare opportunities for communications across disciplinary lines. In universities we say that this is what we like to do, but we seldom have the opportunity to do it.

One minor point of communication: this one has to do with energy demands, how they are treated in different modeling efforts, including that of the CONAES energy modeling resource group and demand panel.

Different elasticities of substitution could be reinterpreted as different code words for: "How easy is it to engage in hard or soft energy technology paths?" I suspect that Amory Lovins could easily find on this questionnaire a set of probabilities that he could assign. That is, he would assign a very high probability to the higher of these two elasticities of substitution; similarly, he would assign a high value to "X," the value of the GNP that he would be willing to give up, to avoid the additional proliferation and diversion impacts.

Indeed, that is the whole trick in designing of such questionnaires. Have we spanned the views that could range from Amory Lovins' to utility executives'? If we try to inform each other of our respective disciplines' code words, we will improve our understanding of this debate.

Now, I would like to ask Dr. Weinberg, who has thought very hard about the different disciplines, about the legitimacy of the techno-economic model as distinct from the ethics-changing viewpoint?

Alvin M. Weinberg: I have a basic sympathy with what Dr. Lindberg said. An acquaintance of mine is an economist, and he suggested to me that the worst thing that happened to the science of economics was the establishment of a Nobel prize in economics. (Actually, I think it was Lady Robinson who originally said that.) It imposes a paradigm on economics; namely, economics is what you can decide . . . what people ought to get a Nobel prize for. So from this point of view, I have a very great sympathy with the kinds of criticisms he offered towards Dr. Manne's and Dr. Richels' analysis.

Dr. Lindberg quoted very extensively from the Bupp and Derian book. Having been involved with the pressurized water reactor a long time ago, I was interested to see what they said. The fact is that for those who have really been in the business in detail and know exactly what happened there, much of what is said in the book has to be taken *cum grano salis.* Derian, for example, says that nuclear energy based on pressurized water is a debacle. But you cannot deny the fact that 12 percent of our energy is coming from nuclear energy. Despite what David Comey has been saying, the operational history of the nuclear reactor is getting better, rather than worse, which is kind of the wrong way, as far as critics of nuclear energy are concerned.

There is one point that Dr. Roberts made with which I have enormous sympathy, and which I paraphrase: "Scientists should be modest." In so much of the discussion that we hear nowadays, where the issues are so nearly transscientific, scientists must require different dimensions of social responsibility. This is illustrated by a discussion I had with Barry Commoner about four months ago, when he first came out with his statement that McDonald hamburgers are going to cause cancer. I asked him, if I ate one hamburger a day, what would the carcinogenic risk be? He said that would be difficult to estimate, but he would estimate it like one cigarette per week. I said, "Under the circumstances, do you think you should have published your report on the carginogenic effects of McDonald's fried hamburgers (Hardie's hamburgers are O.K.) without having done additional experiments that would point out the order of magnitude of the risk?" His answer was, "But you don't want me to withhold the facts, do you?"

Scientists have a responsibility, not only to blow the whistle, but also to say what the limitations are on the knowledge on which they base their whistle-blowing. This is a very important other dimension of scientific responsibility.

The final point I would like to make summarizes the issues. If you look at the whole nuclear debate, you can divide the participants into those who really believe that nuclear energy is an abomination and must be extirpated; for them, the only kind of good nuclear energy is a dead nuclear energy! On the other side are those who feel that nuclear energy is, on balance, a good, not a bad, thing.

The real issue is to construct an acceptable nuclear system. Are there institutional, technological, and, if you like, political changes that are feasible, that one sees ways of reaching, that would, in fact, make nuclear energy generally more acceptable? My own feeling is that unless somehow a stronger consensus can be forged around nuclear energy than it now enjoys, then nuclear energy is not going to survive; an energy system that from 30 percent to 50 percent of the public does not like will not be a feasible energy system. I am spending my time trying to figure out ways of constructing an acceptable nuclear system, looking at the institutional, technological, and political changes that have to be made in order to construct an acceptable nuclear future.

Lincoln Gordon: I want to complement what Dr. Weinberg was just saying by an observation that has come to me while working on the nonproliferation aspects of the problem. It has been very troublesome to me, as I watch people struggling with so-called choices, about energy, about what kinds of society we desire, and so on. What they are really wishing is not just that nuclear power be made good by being made dead—their basic desire is that nuclear energy be disinvented!

That is indeed a profound desire for which I have a lot of sympathy. We would all have been better off if the first chain reaction experiments had turned out to be a blind alley. But nuclear energy *does* work. We must be humble before the facts. Atomic energy *is* possible, and, unfortunately, atomic weapons are also possible. Merely wishing for the disinvention of atomic energy is not a very useful approach to the future, either of international relations or of international politics.

René Malès: I have two comments that will reveal my own bias. My bias is different from the rest of the community, basically an intellectual and university community associated with Argonne. In contrast, I come from an industrial background.

I am dismayed by some of Dr. Lindberg's comments that reflect the lack of recognition that action needs to be taken. He describes for us five new social forms and, in his concluding remarks, describes the desire to come to a new species of decisionmaking. If we make no decisions, if we take no actions until we achieve this new design, the cost to society may be catastrophic. I tried to represent, in my remarks yesterday, the cost of this nondecisionmaking.

As for Dr. Beckerman, I am also dismayed that he speaks of a decision as if it could be made in isolation, as if the United States could make this decision alone. The decision factors in the international scene are very different and compelling. We do not have the luxury of making this decision to gather organically grown food with one hand on the plow, and the other hand on the sewage collector, without resigning from the world. I think this is not a possible or acceptable step.

Stephen Beckerman: I could not agree more that the social environment of any polity is every bit as important in constraining its action as its natural environment. I take the point. However, if it is true that U.S. domestic energy policy is constrained by Soviet activities, it is also true that Soviet policy is constrained by U.S. activities. It is hard to find a point at which the social environment is an independent variable.

Kenneth Shepsle: I do not want to compete with Dr. Roberts in telling stories, but there is a brief one with an important point, I think.

An engineer and a mathematician are on a desert island with two coconut trees as the only source of food. The coconuts being very high up on the trees, they agree that one day the engineer will get the coconuts, and the next day the mathematician will, and so on. The engineer devises a very complicated strategy of shinnying up one coconut tree, using a system of pulleys to assist him, and building a bridge over to the other tree, grabbing a coconut, and working his way back down with the retrieved coconut. The next day, the mathematician follows the same technique, goes slightly higher in the tree to get a coconut, puts it exactly where the previous coconut was, and returns to the ground with a smile on his face. "My God, what have you done?" the engineer asks, and the mathematician replies, "Well, I've taken our problem and reduced it to one with a known solution!"

I think that's an interesting comment on how we deal with the question of "How Should We Decide?" Part of the bias or prejudice of Dr. Manne's approach, and one with which I sympathize, is that it is hard to reduce complicated problems to techniques other than those with which we have some kind of practice and familiarity. And he has demonstrated a capacity to be open to the kinds of comments that some of the critics have made by doing a sensitivity analysis with regard to the particular decision models he has made.

I take Dr. Lindberg's comments, in particular, to be suggestions for the kinds of sensitivity analyses that need to be done—varying some underlying environmental parameter to see whether or not decision paths of one sort or other would be followed. Therefore, the technique of trying to reduce problems to those we think we can grapple with is not all that nonsensical, though it does prejudice our orientation to those techniques of which we already have a command.

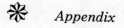 *Appendix*

Post-Symposium Dialogue

Comments by Alvin M. Weinberg
on Papers at Symposium on National
Energy Issues — How Do We Decide?

COMMENTS ON RENÉ H. MALÈS' PAPER
(Chapter 1)

Despite Dr. Malès' implication to the contrary, I believe his analysis does rest rather strongly on his projection of ~140 quads and ~7 × 10^{12} kWh(e) by the year 2000. I would point out that his 140-quad scenario lies on the high side of many current projections, and that he has assumed an increase of productivity ~2 percent per year, which may be on the high side. Our own analysis at the Institute for Energy Analysis, based on a productivity increase of 1.7 percent per year to 2.4 percent per year, leads to a total energy projection of 101 to 126 quads in the year 2000. The point is that, as Dr. Malès says, the future is uncertain. But this means that one can arrive at estimates of the future energy demand that are considerably lower than his on the basis of assumptions that are at least as plausible as his.

I agree with Dr. Malès' underlying contention, that the cost of undercapacity is greater than the cost of overcapacity, up to a point. Obviously if the overcapacity is, say, two or three times the demand, this proposition is no longer sustainable. It therefore does make a difference, whether we project a 140-quad scenario—which seems too high to me—or a 101- to 126-quad scenario. I would urge, therefore, that before we accept Dr. Malès' arguments without qualification, we reexamine the plausibility of his underlying demographic and economic assumptions.

COMMENTS ON HERMAN E. DALY'S PAPER
(Chapter 2)

I want to congratulate Dr. Daly on his beautifully literate statement of the neo-Malthusian position. Although he and Georgescu-Roegen (whom he quotes) are correct in principle in their analysis of the inevitable entropy dilemma, they are wrong in their details insofar as they apply the "entropy trap" to nonenergy resources, and, it seems to me, came to the wrong conclusion in respect to energy. My reasons for so broad an assertion are presented in Goeller and Weinberg (1976), in which we point out that of all mineral molecules man removes from the earth for his purposes, reduced carbon and hydrogen ("CH_x"), that is, fossil fuel, accounts for 80 percent. On the other hand, CH_x is a rare constituent of the earth's crust—about as abundant as neodymium. Thus Dr. Daly and Georgescu-Roegen are correct in their assertion that we shall need an infinite or renewable energy source if we are to survive. They are wrong in implying that only the sun can provide this inexhaustible source. Because breeders can use, as fuel, uranium in seawater or in common rock, the breeder is, for all practical purposes, an infinite energy source also.

Dr. Daly errs, as does Georgescu-Roegen, in lumping all other resources together and implying that they are exhaustible. As far as metals are concerned, we live in the age of iron and aluminum: almost 95 percent of *all* metallic atoms man uses are iron and aluminum. Moreover, those, unlike CH_x, are the two most abundant metals in the earth's crust. We shall *never* run out of them, even in concentrations that are comfortably workable—such as the essentially infinite laterites for iron, clays for aluminum. Thus it is absurd to assert as does Meadows in the *Limits to Growth* (1972) that we shall run out of aluminum in the next thirty years or so.

To be sure, important elements such as copper and zinc are *not* abundant; I suspect when Dr. Daly speaks of the earth's finitude, he has these mainly in mind. But these "scarce" metals represent but 5 percent of man's average "molecule of metal" and 20 percent of its cost. Should the price of "scarce" metals increase tenfold, the cost of an average molecule of metal would about double, *assuming no substitution*. But the opportunity for substitution is enormous. For example, copper can be replaced almost entirely by aluminum. Indeed, the Age of Substitutability would be an age of iron, aluminum, titanium, glass, cement, plastics, and wood—all of which are essentially inexhaustible, provided we have an essentially inexhaustible source of energy. My quarrel with Dr. Daly is not that there are *no* limits; it is rather that the limits are much larger than he implies.

His underlying thesis, that we must turn to the sun alone because it is the only essentially infinite energy source, cannot be taken seriously since the breeder is also an essentially infinite energy source, which, when properly handled, is benign.

COMMENTS ON KENNETH SHEPSLE'S PAPER
(Chapter 4)

Dr. Shepsle takes a rather cynical, if accurate view of the political process in a democracy—that we cannot ever do very much more than achieve small improvements because the political process is riven by parochial interests, desire for short-term gain, and inability to identify the *public* good. I tend to agree with him. It is on this account that I have argued that technical fixes—that is, technical measures that improve the human condition without requiring vast and politically infeasible social changes—are often the only practical way. This is not to say that technical fixes are without deleterious side effects; on the other hand, as Dr. Shepsle would probably agree (given his refreshing academic cynicism) so do social fixes have unforeseen and deleterious side effects. Perhaps we ought to have not only an Office of Technology Assessment, but a companion office of Social Engineering Assessment.

COMMENTS ON CHARLES W. MAYS' PAPER
(Chapter 9)

I must object to Dr. Mays' contention that his data prove linearity at low dose since he has not applied corrections for competing risks. What we seek is the increase in cancer risk caused by radiation and by radiation alone. Thus the number of cancers that appear in the exposed population depends upon how strongly competing causes of death—infection, and so on—intervene. For example, the observed incidence of cancer is an underestimate of the incidence in the absence of competing causes of death, since some individuals who would have succumbed to cancer, in fact, do not do so because they are plucked off by infection, and so on. I would agree with his statement that the possibility of a linear dose-response cannot be rejected by present data, although the final incidence might give a better fit to a concave downwards, or conversely, a concave upwards relation.

I can give a plausibility argument against low-level linearity, and what seems to me to be compelling argument against the $\sqrt{D}$ behavior (concave downward), provided one accepts the existence of thresholds. The individuals exposed to an insult are not homogeneous; their susceptibility to an insult like $f(D)$; that is, $f(D)dD$ is the fraction of individuals whose threshold for injury lies between D and $D + dD$. This I call the "distributed threshold" model. The distribution of susceptibilities is related to $R(D)$, the number who succumb to a dose D, by

$$(1) \qquad R(D) = \int_0^D f(D)dD$$

or

$$(2) \qquad f(D) = R'(D)$$

A linear response at the origin means, according to (2), that there are some individuals who have zero threshold—that is, would succumb to an insult, *no matter how small*. One would then ask why have these individuals not already succumbed to the various background insults? Thus *strict* linearity all the way to the origin is hard to understand, but not quite impossible (since the distributed threshold model may not be correct). On the other hand, if $R(D) \sim D^{\frac{1}{2}}$ (curve downward), then by (2) $f(D) \sim D^{-\frac{1}{2}}$—i.e., $f(D)$ is infinite at the origin. This is too bizarre an implication of $R(D) \sim D^{\frac{1}{2}}$ to allow me to accept $D^{\frac{1}{2}}$ at the origin. Of course, as Dr. Mays points out, there are other analytic forms that are concave downward at the origin, but do not suffer this deficiency.

COMMENTS ON PAPERS BY
NORMAN RASMUSSEN (Chapter 5)
AND GORDON BURLEY (Chapter 6)

I congratulate Dr. Rasmussen for introducing explicitly what I call a "relational" metaphysic for establishing standards of acceptable risk. By "relational" metaphysic I mean the idea of relating risks to other inevitable or at least generally accepted risks. H. Adler and I have suggested (1978) that low LET radiation be taken as the standard against which other carcinogens are judged—that is, we universalize standards of risk by comparing them to the low LET standard. The latter we set equal to the standard deviation of the natural background. This is about 20 mrem per year, essentially the EPA standard for the nuclear fuel cycle; Dr. Burley had showed that this was the standard deviation of the natural background several years ago. The same suggestion appears in the American Physical Society Study on the Nuclear Fuel Cycle and Waste Disposal.

For comparing risks, we have proposed that the linear hypothesis be used in what I call its "weak" form: we use it to *compare* risks, but not to estimate absolutely the number of casualties caused by an insult. I hope that Dr. Rasmussen's excellent paper encourages others to examine such *relational* approaches to universalizing standards of risk.

COMMENTS ON LEON N. LINDBERG'S PAPER (Chapter 16)

I find much in Dr. Lindberg's argument with which to sympathize; in particular, his plea to temper econometrics with behavioral insights. Yet, I believe his beautifully articulate presentation illustrates all too clearly the difficulties we face in discussions between technologists and behavioral scientists.

Dr. Lindberg seems to have in his mind that plutonium, as well as nuclear energy, is bad and must be extirpated. (If this is not his view, I hope he corrects me.) Moreover, he seems to believe that the deficiencies of nuclear energy cannot be corrected. He apparently has come to this view by study of the nuclear debate; and, at least by implication, he concludes that the arguments of those who oppose nuclear energy are sounder than those of the proponents.

In his presentation, however, he quotes extensively from Bupp and Derian. I would insist that Bupp and Derian simply cannot be used as authorities on which to base an estimate of the state of the nuclear enterprise. Their main point seems to be that light water reactors are much more expensive than optimists like me thought they would be, and that they are much less safe than Dr. Rasmussen contends them to be. To base his position on this authority is dangerous; for example, despite the claims in Bupp and Derian that light water reactors are thrashing about in a death throe, 12 percent of our electricity came from LWRs in 1977; and the experience during 1978 was better, rather than worse, than previously. I would urge that Dr. Lindberg read Rossin (1978) before concluding that Bupp and Derian represent the most knowledgeable or dispassionate assessment of light water technologies. Thus, as is so common in the bitterly polarized nuclear debate, we find participants in the debate selectively quoting authorities without subjecting the assertions of those authorities to searching scientific review.

This brings me to my final point, one brought out by Dr. Roberts. The issues we deal with are beclouded by uncertainty; many of them are probably transscientific. Under the circumstances all of us—technologists and nontechnologists—ought to be careful to state the limits of our knowledge, and we should, in discourse such as we have

had in this symposium, be prepared to expose our degree of ignorance both in the fields in which we claim no expertise and in those
in which we do.

REFERENCES

Adler, H.I. and A.M. Weinberg. 1978. "An Approach to Setting Radiation
Standards." *Health Physics* 34 (June): 719–20.

Goeller, H.E., and A.M. Weinberg. 1976. "The Age of Substitutability."
Science 191 (February 20): 683–89.

Meadows, D.H., D.L. Meadows; J. Randers; and W.W. Behrens, III. 1972.
Limits to Growth. New York: Universe Books.

Rossin, A.D. 1978. "Economics of Nuclear Power." *Science* 201 (August 18): 582–89.

A Response to
Alvin M. Weinberg's Comments

Kenneth Shepsle

Dr. Weinberg's collection of comments is balanced, insightful, and sensitive. I am happy he finds in my own remarks a "refreshing academic cynicism." The position I take is that most democratic polities are confronted with a congeries of interests (in the energy area or any other). There is no morally superior "public interest"; rather, there are only self-anointed, morally superior individuals who proclaim their preferences are in the "public interest." We must learn to look below the surface of those proclamations. Once we do, we will not necessarily discover narrow, selfish, self-serving preferences (a belief that a truly cynical person would hold); but rather, we will find personal preferences, beliefs, and opinions—some selfish, some altruistic. The important word to emphasize here is "personal," for that is what all these beliefs, opinions, and preferences are.

Actually, while he agrees with me, Dr. Weinberg's remarks illustrate, I believe, some of the problems the layperson has in evaluating expert opinion; these problems are produced by the interaction of *expertise* and *interest*. To the outsider it would not come as any great surprise to find an eminent scientist and science administrator arguing in behalf of "technical fixes." The more cynical outsider (and I plead innocent to that charge) would sense self-interest lurking below the surface. The less cynical would perceive a more enlightened quality, as in the genuine faith of a scientist in "technical measures that improve the human condition without requiring vast and politically infeasible social changes."

The bottom line, for me, I suppose, in regard to how we should evaluate evidence and argument (and, ultimately, how we should decide) on vital matters such as the future production of energy, is to be dubious of ideologues, both political and scientific, for ideologues tend to be unreceptive to information inconsistent with their world views. It is not so much that their minds are made up (I am not talking about "true believers") as it is that their minds are often incapable of being changed. The range of frequencies of their data receivers is restricted to a narrow portion of the band. My dubiousness, however, does not reflect an underlying moral judgment. (I am not, to repeat, a cynic.) It is, rather, a practical strategy by which I evaluate competing claims. That is, each claim is a reflection of a private interest or a private belief about what is in the "public interest." Thus, in my view, all articulators of claims are special pleaders, whether they represent the AAAS, the Clamshell Alliance, the Nuclear Regulatory Commission, or, for that matter, Argonne National Laboratory. Some special pleadings can be ignored out of hand, if they lack even the pretense of supporting logic or evidence. But many such pleadings cannot be so easily dismissed, even though, as a collection, they may be at odds with one another. And it is at this point that we ought not limit our alternatives in any nonreversible fashion. While we need not let a thousand flowers bloom, we should permit six or seven varieties along with placing some priority on superintending the garden. In this light, it is in his acknowledgment of uncertainty and of the limits on expertise that Dr. Weinberg's comments are so appropriate. His final paragraph bears rereading and remembering.

Response by Charles W. Mays
to Comments by Alvin M. Weinberg
on Shape of the Dose-Response Curve

Dr. Weinberg raises some interesting points on the shape of the dose-response curve for bone sarcoma induction by low doses of plutonium-239 in beagles, based on our data, as analyzed by Peter Groer (1978). This study is still in progress and involves many dogs originally intended for sequential sacrifice. However, only a few were sacrificed and the vast majority were redesignated for lifespan toxicity studies (Mays et al. 1979). I publicly apologize to Dr. Groer that the preface to our injection tables neglected to mention this important redesignation and that, as a consequence, Dr. Groer's paper omits many dogs that should have been included. It is particularly unfortunate that Dr. Groer's analysis omitted the dog with bone sarcoma at an average skeletal dose of 2 rad, and included only one of the three dogs with bone sarcoma at 22 rad.

For the reader's convenience I give the status as of 31 March, 1979 for our dogs with low-level plutonium (Table A−1 and Figure A−1) and, also, for our dogs with low-level radium (Table A−2 and Figure A−2). Under the classification bone sarcomas, we include osteosarcomas, chondrosarcomas, and fibrosarcomas. The average survival time of the dogs receiving low-level radiation is about eleven years after injection. Allowing for a tumor growth period of one year, the most important dose is received during the first ten years. When all of the dogs have died, the fraction with bone sarcomas will lie between the indicated limits. The lower limit would be reached if none of the living dogs develop bone sarcomas, whereas if all of the living dogs develop bone sarcomas, the upper limit would be attained. As the dogs continue to die, these limits will converge

Table A–1. Bone Sarcomas in Low Dose 239Pu Beagles *(31 March 1979)*

Injected Dogs (μCi ^{239}Pu/kg)	Injected Dogs	Living Dogs	Dead Dogs	Sarcoma Dogs	Sar. Dogs / Dead Dogs (%)	Av. Skel. Dose 10 Year Post Inj. (rad)
0.016	26	10	16	5	31	57
0.010	38	35	3	1	33	37
0.0055	38	27	11	3	27	20
0.0018	46	21	25	0	0	7
0.00070	28	15	13	1	8	3
Control[a]	133	35	98	1	1	0

[a]Young adult controls for ^{239}Pu, ^{226}Ra, ^{228}Ra, ^{228}Th, and ^{90}Sr.

Figure A–1. Bone Sarcoma Incidence in Young Adult Beagles at Low Doses from [239]Pu. The ratio of bone sarcoma dogs/dead dogs, as of 31 March 1979, is plotted as a solid circle for each dose level. The final incidences, when all of the dogs have died, will be between the indicated upper and lower limits. Shown for comparison is the linear slope from our previously completed study in 28 beagles with plutonium doses from 55 to 135 rad, in which 14 developed bone sarcomas (Mays et al 1976). In the on-going low dose plutonium study, the possibility of a linear dose-response cannot be rejected by present data, although the final incidence might give a better fit to a concave downwards, or conversely, a concave upwards relationship.

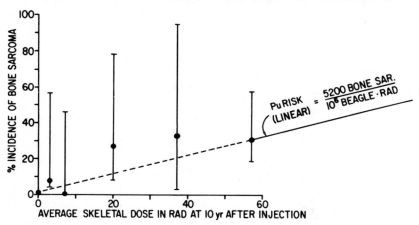

BONE SARCOMA INCIDENCE IN BEAGLES AT LOW DOSES FROM [239]Pu (31 Mar. 1979)

together at the final incidence. Since most dogs at the low dose levels are expected to die without bone sarcomas, the final incidence should, in general, be closer to the lower limit than to the upper.

For each dose level, the present ratio of dogs with bone sarcomas to dead dogs is shown as a crude approximation of what the final incidence might be. Our previous experience with low doses that produce delayed tumor appearance times is that the ratio of dogs with sarcomas to dead dogs tends to increase somewhat as the dogs continue to die. Thus, preliminary results, which sometimes suggest a concave upwards response for α-particle induction of bone sarcomas (Mays and Lloyd 1972) may change into a more linear response when the full results have been obtained (for example, compare our preliminary results for thorium–228 in beagles (Mays and Lloyd 1972) with the final results in Figure 9–6 of my paper at this symposium).

Table A–2. Bone Sarcomas in Low Dose ^{226}Ra Beagles (31 March 1979)

Injected (μCi ^{226}Ra/kg)	Injected Dogs	Living Dogs	Dead Dogs	Sarcoma Dogs	Sarcoma Dead Dogs (%)	Av. Skel. Dose 10 Year Post Inj. (rad)
0.062	23	3	20	2	10	210
0.022	25	7	18	1	6	74
0.0074	10	2	8	0	0	25
Control[a]	133	35	98	1	1	0

[a]Young adult controls for ^{239}Pu, ^{226}Ra, ^{228}Ra, ^{228}Th, and ^{90}Sr.

Figure A—2. Bone Sarcoma Incidence in Young Adult Beagles at Low Doses from ^{226}Ra. The ratio of bone sarcoma dogs/dead dogs, as of 31 March 1979, is plotted as a solid circle for each dose level. The final incidences, when all of the dogs have died, will be between the indicated upper and lower limits. Shown for comparison is the linear slope from our previously completed study in 51 beagles with radium doses from 183 to 2500 rad, in which 17 developed bone sarcomas (Mays et al 1976). In the on-going low dose radium study, the possibility of a linear dose response cannot be rejected by present data, although the final incidences might give a better fit to an alternative relationship.

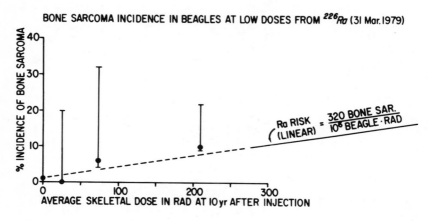

BONE SARCOMA INCIDENCE IN BEAGLES AT LOW DOSES FROM ^{226}Ra (31 Mar.1979)

In view of all these considerations, I conclude that the possibility of linearity cannot be rejected by our present data on bone sarcoma induction by low doses of plutonium–239 and radium–226 in beagles (see Figures A–1 and A–2). This does *not* prove that the dose response is exactly linear, and when this experiment is finished the final incidence may best support a concave upwards, or conversely, a concave downwards response. But these final results will take several more years and cannot be hurried in time despite the urgency of our impatience.

Dr. Weinberg is quite right that the incidence (Y) cannot be proportional to the square root of dose $\sqrt{D}$ over the entire dose range because of the infinite slope it predicts at the origin. But there are other models, such as $Y = aDe^{-kD}$, that also lead to a concave downwards dose-response. This particular example has a finite slope (a) at the origin. Furthermore, it has an exponential factor (e^{-kD}) in the dose response to allow for cell-killing by high LET radiation. One can conceive of many other models that are concave downwards. Mathematically disproving only one of the many equations that are concave downwards does not necessarily disprove the possible validity of the others.

A linear response at the origin does *not* mean "that there are some individuals who are completely susceptible—i.e., would succumb to an insult, *no matter how small.*" Quite the contrary, the linear response means that the chance of an individual being affected approaches zero as the dose approaches zero (if $Y = aD$, then $Y \to 0$ as $D \to 0$).

With regard to a possible "sigmoid" appearance of curves, it is very important that both the incidence and the dose be plotted on linear scales. Robley Evans et al. (1969) have clearly shown that the linear nonthreshold response is distorted into a curve that is strongly concave upwards when the incidence is plotted on a linear scale and the dose is plotted logarithmically.

The Kaplan-Meier method (1958) is a powerful tool for correction of competing risks. I have used it myself to show that the mammary tumors in our colony are not appreciably induced or repressed by bone-seeking radionuclides (Taylor et al. 1976); that the life-shortening in our dogs at medium and low doses is accounted for by radiation-induced cancers, primarily bone sarcomas (Mays and Dougherty 1972); and that the rate of leukemia induced by phosphorous–32 in patients with polycythemia vera is in excess of that associated with their untreated disease (Mays 1973). But Kaplan-Meier analysis has its limitations. It always gives a final cumulative incidence of 100 percent whenever the last animal in a group develops the disease of interest. For example, if in 100 animals only the last dies of osteosarcoma, the cumulative incidence is zero up to this time and 100 percent thereafter. Both of these values, especially the latter, could be very misleading. The conventional 1 percent incidence would seem much more useful. Kaplan-Meier statistics translate from the real world, where a multitude of competing risks exist, into an artificial world where only the disease of interest is considered. This results in calculated cumulative incidences that usually exceed the real incidences, because in the real world many people die from competing risks before they can develop the disease of particular interest to the investigator. So if risk coefficients derived from Kaplan-Meier statistics are to be applied to actual human populations, proper adjustments must be made.

I have used the conventional incidence (fraction of dogs with bone sarcomas) because: (1) it is a simple concept easily explained and understood; (2) bone sarcomas are the main radiation-induced cause of death in young adult beagles given bone-seeking radionuclides

(Mays and Dougherty 1972), so that the radiation produces few other competing deaths; and (3) most of our dogs with low-level radiation have long lifespans and die from diseases of old age, so that the competing risk of death from nonradiation cause is rather similar among the low-dose levels.

Of course, Kaplan-Meier analysis should also be used as an independent method of analysis, but with a clear understanding of its limitations. For example, it does not include the effect of tumors that have not yet appeared. Pete Groer (1978) understands this completely, for the footnote to his Table 1 states, "It has to be emphasized that, in the lower dose groups, most beagles are still alive and that some osteosarcomas may occur during the next several years. Such additional tumors will make the corresponding incidence greater than zero."

Thus, it may be premature at this time (and perhaps even wrong) to conclude that the dose-response is "quadratic" for bone sarcoma induction by low doses of plutonium–239 in beagles.

REFERENCES

Evans, R.D.; A.T. Keane; R.J. Kolenkow; W.R. Neal; and M.M. Shanahan. 1969. "Radiogenic Tumors in the Radium and Mesothorium Cases Studied at M.I.T." In *Delays Effects of Bone-Seeking Radionuclides*, C.W. Mays, W.S.S. Jee, R.D. Lloyd, B.J. Stover, J.H. Dougherty, and G.N. Taylor, eds., pp. 157–94. Salt Lake City: University of Utah Press.

Groer, P.G. 1978. "Dose-Response Curves and Competing Risks." *Proc. Natl. Acad. Sci.* (September): 4087–91.

Kaplan, E.L., and P. Meier. 1958. "Nonparametric Estimation from Incomplete Observations." *J. Am. Statist. Assoc.* 53: 457–81.

C.W. Mays. 1973. "Cancer Induction in Man from Internal Radioactivity." *Health Phys.* 25: 585–92.

Mays, C.W., and T.F. Dougherty. 1972. "Progress in the Beagle Studies at the University of Utah." *Health Phys.* 22: 793–801.

Mays, C.W., and R.D. Lloyd. 1972. "Bone Sarcoma Incidence vs. Alpha Particle Dose." In *Radiobiology of Plutonium*, B.J. Stover and W.S.S. Jee, eds., pp. 409–30. Salt Lake City: J.W. Press, University of Utah.

Mays, C.W.; H. Spiess; G.N. Taylor; R.D. Lloyd; W.S.S. Jee; S.S. McFarland; D.H. Taysum; T.W. Brammer; D. Brammer; and T.A. Pollard. 1976. "Estimated Risk to Human Bone from [239]Pu." In *Health Effects of Plutonium and Radium*, W.S.S. Jee, ed., pp. 343–62. Salt Lake City: The J.W. Press, University of Utah.

Mays, C.W., G.N. Taylor, W. Stevens, W.S.S. Jee, and M.E. Wrenn. 1979. "Bone Sarcomas at Low Doses of α-Radiation in Beagles." In *Research in Radiobiology*. University of Utah Report COO–119–254, pp. 9–12.

Taylor, G.N.; L. Shabestari; J. Williams; C.W. Mays; W. Angus; and S. McFarland. 1976. "Mammary Neoplasia in a Closed Beagle Colony." *Cancer Res.* 36: 2740–43.

Comments by Peter G. Groer
on the Response by Charles W. Mays
to Comments by Alvin M. Weinberg
on the Shape of the Dose-Response Curve

I was not present at the meeting but am offering these comments since Dr. Mays has criticized some aspects of my published analysis (1978) of the data on plutonium–239 in beagles. I analyzed the occurrence of *osteosarcomas* not of *bone sarcomas* (see Dr. Mays' response for a definition of bone sarcomas). This is clearly stated in my paper. Osteosarcomas should not be analyzed together with the other bone sarcomas because the mechanism of induction is different. The osteosarcomas, incidence increases with dose, but no such trend is noticeable for the incidence of chondro- and fibro- sarcomas. The bone sarcoma at 2 rad and one of the sarcomas at 22 rad are chondrosarcomas and were therefore omitted from my analysis. The osteosarcoma at 22 rad, which occurred in a special-assignment dog, should be included after Dr. Mays' explanation of the redesignation of many special-assignment dogs. If this osteosarcoma is taken into account, the incidence as estimated by the Kaplan-Meier procedure *decreases* from 25 to 18 percent. I agree with Dr. Mays that the possibility of a linear dose-response for plutonium–239 in beagles cannot be rejected at the present time. However, a sigmoid response also cannot be rejected. Dr. Mays' point about the distortions arising from the use of semi-log plots is correct, and an amendment to my paper (1978) that clarifies this point exactly will be published to avoid possible misinterpretation.

Reasons that the Kaplan-Meier (K-M) estimator should be used to estimate incidence in a competing risk situation are given in my reference 1 and I will not repeat them here. In one example, Dr. Mays points out that he used the K-M procedure to show that there is an excess of leukemias in patients treated with phosphorous–32. Why should this procedure not be used to show that there are excess osteosarcomas in beagles treated with plutonium–239?

It is correct that the cumulative incidence is 100 percent if only the last out of 100 animals develops the disease of interest. This simply means that there are not enough events to obtain a reliable estimate of the distribution function, and better data from a larger group of animals are needed. But I agree that a blind application of the K–M estimator is misleading in this situation. This is, however, a rare situation, and the "pathologic" behavior of the estimator should remind the investigator that more data are needed.

My arguments against the use of the conventional incidence can be found in my paper (1978). It is clear that "adjustments" have to be made if K–M risk estimates from one exposed population are to be used for other populations. But this is the correct way of transferring risk estimates. The K–M estimator gives the best possible estimate, free of the influence from competing causes of death in the exposed population. It should be used to calculate the "actual" risks in another population with a different competing risk structure. The K–M procedure like every other statistical technique has no prophetic power and can therefore not include the effects of tumors that have not yet appeared.

REFERENCE

Groer, P.G. 1978. "Dose Response Curves and Competing Risks." *Proceedings of the National Academy of Science* 75: 4087–91.

Reply to Alvin Weinberg's Post-symposium Comments

Leon N. Lindberg

I am afraid that I did not make my point clearly enough. I explicitly did *not* take the position that nuclear energy was bad and that its deficiencies cannot be corrected. The behavioral scientist can find as many searching scientific arguments on either side of the many contentious issues within the nuclear debate, and it was not my purpose to line up with any of them.

The quotations from Bupp and Derian (1978), who are, I repeat, generally favorable to nuclear energy on technological grounds, were intended to illustrate one point only. That was that the political choice and evaluation processes revealed by the history of nuclear development were unacceptable to me on grounds of democratic theory and practice. Otway and Thomas (1977) make a similar point when they write:

> technologists have been surprised at the strength of public reactions against nuclear energy. Social scientists, in turn, have been equally surprised at the reluctance of technologists to accept the validity of public concerns about social issues, as being quite distinct from the technical realities . . . if there is a central issue in the nuclear controversy, it is personal and political power, and public participation in the control of that power . . .

REFERENCES

Bupp, Irvin C., and Jean-Claude Derian. 1978. *Light Water*. New York: Basic Books.

Otway, Harry, and Kerry Thomas. 1977. "Understanding Public Attitudes toward Nuclear Power." Paper prepared for IAEA/IIASA, Vienna, November 22. Stencil.

A Response to
Dr. Weinberg's Comments

Gene Rochlin

Much as I respect Dr. Weinberg's sincerity and reputation, his summary is more of a tangent, carrying us away from the matters that were dealt with directly and into other areas. I do not deny that these are related, but a full and comprehensive discussion of them would take much more time and space than I am able to give to the them now. Nevertheless, let me sketch out at least a few points on which I appear to be in considerable disagreement with the views Dr. Weinberg presented.

In his response to Dr. Daly, Dr. Weinberg states that the breeder is also an (almost) infinite energy resource, if seawater or common rock is used as a source of uranium. The assumption is, of course, that the breeder will make these economically available as resources. But it is not sure that even breeder reactor economics will make these dispersed resources economically recoverable, nor is it clear what the energy and environmental costs of their recovery will be. The comparison is too offhand, and too reminiscent of the early days of LWR development, when advocates were claiming that nuclear power would be so cheap it would be practically given away. In any case, that was not the main point of Dr. Daly's argument at all. As with many papers that argue the merits of the "soft path," it was based on social and political, as much as technical considerations. There are many who would consider a society based on the plutonium-fueled fast-breeder to be less desirable than one based on solar energy, even if there were millions of years of uranium ore very accessible and "minable." The solar versus breeder argument does not turn on the issue of long-term energy supply alone.

With regard to the use of abundant energy to avoid resource scarcity, I will do no more than summarize the more general criticism of this view. Iron and aluminum are very abundant in the crust, but the limit on their extraction is set by not only the energy, but the entropy and environmental costs of extracting them, unless they are present in relatively concentrated deposits. Moreover, neither pure iron nor pure aluminum are of very great use. Limitations on primary metal resources as *usable* for sophisticated technical and industrial purposes are more likely to be set by the scarcer ores for cobalt, chromium, molybdenum, nickel and so on, with which they must be alloyed to obtain desirable properties such as hardness, stiffness, and resistance to corrosion.

The "age of substitutability" of which Dr. Weinberg apparently dreams, is not, to many of us, either a realizable or a desirable idea. The costs of its achievement may simply be too great in social, political, or environmental terms. I am reminded of the simple aphorism of the Indians of the California foothills, who were driven from their land by a shortage of their primary source of meal—the pine nuts from digger pines. "The Indian," they said, "harvested only the pine cones that fell, and sometimes went hungry. The white man, however, cut the pines to harvest many nuts—and now we always go hungry." There are many analysts of the energy situation who believe that we cannot go on with bigger, more elaborate, more socially complex, more politically-trying energy technologies, in an attempt continually to defy the finiteness of the earth and the growth of entropy. For many, adjustment to the type of sustaining society characterized above by the "Indian" does not mean going back to the woods or to any other anthropologically primitive stage, but learning to live on and with the earth and natural energy flows by means of high and sophisticated but smaller-scale and "appropriate" technology, rather than large-scale and "defiant" technology.

Dr. Shepsle in his presentation was trying to point out that neither technical nor social fixes address the substantive political and social problems presented by technology. The political system is inherently parochial and full of special-interest pleading. But what are the alternatives? The point is not to abandon hope for the system, nor to overturn it, but to recognize that disinterested political behavior for the safe or reasonable development of alternatives is to some extent incompatible with political realities. This is not news. As for the public good, I have some disagreement both with Dr. Shepsle's presentation and with Dr. Weinberg's interpretation that it is so elusive. Since the moral and ethical debate as to what is the public good has raged at least since the time of Plato, I am a bit reluctant to enter into it with any simplified statement. However, it is noted that, whatever the "absolute" public good may be, populaces and societies have not had all that much difficulty making relative judgments.

Dr. Weinberg appears to be chastising Dr. Lindberg for having come to certain conclusions by "study of the nuclear debate," certainly an unprovable and, I believe, an incorrect assumption. The case of Bupp and Derian's study is to me a classic example of assuming that a critic must be either an enemy or an opponent, a mistake made so often in the aforementioned nuclear debate as to characterize it sociologically. A careful reading of the book of Bupp and Derian in no way supports the frequent accusation that their views are "antinuclear." Because his argument has been picked up by nuclear opponents who contend that nuclear power is uneconomic, somehow Bupp is held to be among those who oppose the further use of nuclear power—which is certainly not true. That he has made a persuasive case for the *abuse* of a technology within his particular expertise in business and economics is simply overlooked in the race to divide all participants in the debate into two camps.

At the risk of seeming to act in exactly the fashion for which I criticized Dr. Weinberg, I assert that the article by Rossin that he cites to rebut the work of those such as Bupp and Derian is simply irrelevant. It may be true that Commonwealth Edison, using plants bought in the days when vendors were taking a loss to promote sales, may be obtaining an advantage from them. This is precisely the type of case Bupp and Derian cite as having led to overoptimism. In the real and present world of hard cash and tight money, coal and nuclear are running neck and neck, and the differential between them is, in the minds of purchasers, less than the cost of uncertainty in either. The fairest statement to make at the present is that one cannot tell for certain which would prove the most economical over the next thirty years (the life of the plant).

As for Dr. Weinberg's final point, I am very cautious about endorsing it despite its surface impeccability. All too often this type of statement is twisted to argue that only the so-called hard facts can be admitted to the debate. To use a metaphor, I can apply any number of hard facts to choose between one egg and another for breakfast, but no set of facts will inform me if I am trying to choose between an egg and a bowl of cereal. It has taken many years to get the DOE and its predecessors to admit to their own technical doubts about so-called hard facts; witness the recent spate of reports on evaluation of geological environments for nuclear waste disposal. Only under terrific social and political pressure did the real uncertainties in the evaluations underlying previous choices for disposal sites and media become generally known. When acting in ignorance, one must proceed with caution. To deny ignorance and uncertainty is to encourage proceeding recklessly and without warning of possible pitfalls. There is an understandable unease in dealing with "soft" facts among those with technical and scientific training, and an equally understandable desire among politicians and public alike to have some clear-cut and unambiguous criterion for making a difficult decision. This is precisely why every effort must be made to resist the siren song of certainty where no real certainty can exist, and to work as best we can to grapple with the necessity to make difficult choices on uncertain grounds with at best partial data. To do so requires the most judicious balancing of optimism and pessimism about the outcome of our actions, a judgment that can be constructed only if they are seen for what they are—primarily political, rather than technical choices.

Index

Roberts, M.J., 64-65, 100-101, 144,
  208
  "National energy decisionmaking:
    rationalism and rationalization,"
    211-240
  comments by S. Beckerman,
    288-289
  comments by H. Daly, 67
  comments by L.N. Lindberg, 270,
    272-274
  comments by R. Malès, 68
  comments by M.B. Smith, 278-279
  comments by A.M. Weinberg, 294
Rochlin, G., 63-64, 146-147, 149,
  208
  "Sociopolitical point of view,"
    87-92
  discussion, 93-94, 97, 100
  comments by S. Beckerman, 99,
    288
  response to "Comments" by A.M.
    Weinberg, 318-321
Rocky Flats plant, 84
Rousseau, J.-J., 212, 221, 225, 229,
  288
Rowen, H.S.
  "Nonproliferation criteria for
    assessing civilian nuclear tech-
    nologies," 179-194
  comments by L. Gordon, 195-200
  comments by B.D. Zablocki,
    205-206
Rowland, R.E.
  "Some comments on the toxicity
    of plutonium-239," 105-126
  discussion, 143-149
  comments by C.W. Mays, 127, 130,
    138, 140
Ruthenium, 89

Saccharin, 77-78
Safety criteria, 73-82
  aircraft landing systems, 75-76
  chemical plants, 76
  Delaney Amendment, 77-78
  dikes in the Netherlands, 76-77
  nuclear power plants, 78-79
  plutonium, 79-82
SALT, 182
Sampson, A., 287n
Sanders, B.S., 108
Saperstein, A.M., 101-102
Sax, J.L., 229
Say's Law, 46, 67
Schales, F., 135
Schumacher, E.F., 34, 48
Seabrook power plant, 68
Sewage treatment plants, 230

Sewell, W.
  Opening remarks, 152
Shepsle, K.A., 65, 296
  "Economic growth and national
    energy policy: some political
    facts of life," 53-61
  comments by H. Daly, 66-67
  comments by L.N. Lindberg, 270,
    273
  comments by R. Malès, 68
  comments by G. Rochlin, 319
  comments by A.M. Weinberg, 302
  reply, 306-307
Sills, D.L., 72
Sinclair, W., 145
Slobodkin, L., 62
Smith, M.B.
  "Psychological point of view,"
    277-281
Social policy
  and decision theory, 211-240
  government role in, 221-239,
    272-273
  related to energy policy, 51
Soddy, F., 23
Solar biomass power, 290
Solar energy, xvii-xviii, 3, 5, 26-28,
  37, 41, 196, 287, 318
Spiess, H., 134-136
SRI International, 7
Standard of living
  and energy consumption, 62-63
Starr, C., 90, 96
State government
  and social policy, 230-231
Stern, P.M., 287n
Stewart, A., 107-111, 143
Strontium-90, 145
Suppliers' Group, 200

Thermal pollution, 93
Thermal reactors, 157, 159, 164-167,
  177, *158-160*
Thorium, 119-122, 147-148,
  157-177 *passim*, 191, 193, 196,
  253, *159, 175-176*
Thorium-227, 131-132, *132, 137*
Thorium-228, 127, 132, *133, 137*
Thorium-232, 154-155, 164, 173
Thorotrast, 139
Three Mile Island reactor, xx
Till, C.E.,
  "Technical considerations in
    decisions on plutonium use,"
    153-178
  Comments by L. Gordon, 195-196,
    198

Contributors

Stephen Beckerman
Assistant Professor of Anthropology
Southern Methodist University

Gordon Burley, Director
Environmental Standards Branch,
   Office of Radiation Programs
U.S. Environmental Protection Agency

Herman E. Daly
Professor of Economics
Louisiana State University

Lincoln Gordon
Senior Fellow
Resources for the Future

Robert V. Laney
Deputy Laboratory Director,
   Operations
Argonne National Laboratory

Lester B. Lave
Senior Fellow,
Brookings Institute and
Professor of Economics
Carnegie-Mellon University

Leon N. Lindberg
Professor of Political Science
The University of Wisconsin

René H. Malès, Director
Energy Analysis and Environment
   Division
Electric Power Research Institute

Alan S. Manne
Professor of Operations Research
Stanford University

Charles W. Mays
Research Professor of Pharmacology
Radiobiology Division
College of Medicine
University of Utah

Norman C. Rasmussen
Professor and Head of Nuclear
   Engineering Department
Massachusetts Institute of Technology

Richard G. Richels
Energy Systems Program
Electric Power Research Institute

Marc J. Roberts
Professor of Political Economy
  and Health Policy
Harvard School of Public Health

Gene I. Rochlin
Research Policy Analyst
Institute of Governmental Studies
University of California, Berkeley

Henry S. Rowen
Professor of Public Management
Graduate School of Business
Stanford University

Robert E. Rowland, Director
Material Science Division
Radiological and Environmental
  Research Division
Argonne National Laboratory

Robert G. Sachs
Professor of Physics
Enrico Fermi Institute
The University of Chicago

William H. Sewell
Vilas Research Professor of
  Sociology
The University of Wisconsin

Kenneth A. Shepsle
Professor of Political Science
Washington University

David L. Sills
Executive Associate
Social Science Research Council

M. Brewster Smith
Professor of Psychology
University of California, Santa Cruz

Charles E. Till, Director
Applied Physics Division
Argonne National Laboratory

Alvin M. Weinberg, Director
Institute for Energy Analysis
Oak Ridge Associated Universities

Benjamin D. Zablocki
Associate Professor of Sociology
Rutgers University
and Senior Research Associate
Center for Policy Research
New York City

About the Editor

Robert G. Sachs is a theoretical physicist whose research and teaching interests over the years have centered on the fundamental questions concerning the structure of matter, most recently with emphasis on particle physics. He has also served as a scientific administrator of applied science and engineering programs as well as fundamental research programs. He was Division Director (Theoretical Physics) and Associate Laboratory Director for High Energy Physics at Argonne National Laboratory, and Director of The Enrico Fermi Institute of The University of Chicago. When he became Director of Argonne in 1973 the energy debate was just beginning and Argonne, as a major center for the development of energy technology, was thrust into the middle of it. It was during this period, 1973–1979, as Director of Argonne that he became involved with the issues addressed by this symposium. He is presently Professor in the Physics Department and The Enrico Fermi Institute of The University of Chicago.

Sponsors